D0820432

Beatrice Stillman is a writer and translator with a special interest in the situation of women in present and past societies. Among her previous publications was a much acclaimed translation of *Dostoevsky: Reminiscences,* by Anna Dostoevsky, the writer's wife. She is now at work on a full-scale biography of Kovalevskaya.

A
Russian
Childhood

Sofya Kovalevskaya

A Russian Childhood

TRANSLATED, EDITED AND INTRODUCED BY
BEATRICE STILLMAN

With an Analysis of Kovalevskaya's Mathematics
by P. Y. Kochina
USSR Academy of Sciences

Springer-Verlag
New York Heidelberg Berlin

This book has been selected for inclusion in the *Sources and Translation Series of the Russian Institute,* Columbia University.

AMS classification 01A70

Library of Congress Cataloging in Publication Data

Kovalevskaia, Sof'ia Vasil'evna Korvin-Krukovskaia, 1850-1891.
A Russian childhood.

Translation of Vospominaniia detstva.
Bibliography: p.
Includes index.
1. Kovalevskaia, Sof'ia Vasil'evna Korvin-Krukovskaia, 1850-1891. 2. Mathematicians—Russia—Biography. I. Title.
QA29.K67A3513 510'.92'4 [B] 78-12955

Printed in the United States of America.

9 8 7 6 5 4 3 2 1

ISBN 0-387-90348-8 Springer-Verlag New York
ISBN 3-540-90348-8 Springer-Verlag Berlin Heidelberg

For Pelageya Yakovlevna Kochina,
untiring scholar and generous spirit,
in admiration and friendship

Contents

Preface

In the year 1889 Sofya Vasilievna Kovalevskaya, Professor of Mathematics at the University of Stockholm, published her recollections of growing up in mid nineteenth century Russia. Professor Kovalevskaya was already an international celebrity, and partly for the wrong reasons: less as the distinguished mathematician she actually was than as a "mathematical lady"—a bizarre but fascinating phenomenon.*

Her book was an immediate success. She had written it in Russian, but its first publication was a translation into Swedish, the language of her adopted homeland, where it appeared thinly disguised as a novel under the title *From Russian Life: the Rajevski Sisters* (Sonja Kovalevsky. Ur ryska lifvet. Systrarna Rajevski. Heggström, 1889).

In the following year the book came out in Russia in two

---

*"My gifted Mathematical Assistant Mr. Hammond exclaimed . . . 'Why, this is the first handsome mathematical lady I have ever seen!'" Letter to S. V. Kovalevskaya from J. J. Sylvester, Professor of Mathematics, New College, Oxford, Dec. 25, 1886.

installments of the journal *Vestnik Evropy* (Messenger of Europe), in the autobiographical form and the language in which it had been written. It was called *Memories of Childhood.* The editor of another journal, *Russkaya starina,* one Mikhail Ivanovich Semevsky, went so far as to maintain that Kovalevskaya's work was worthy of standing side by side with Tolstoy's *Childhood:* an exaggerated but understandable tribute to a native daughter from a public figure who, twenty-seven years before, had presumed to propose marriage to her elder sister Anyuta and been rejected by their father as a penniless upstart.

The success of her book in her native land was to be one of the last sources of pleasure in Kovalevskaya's life, for within six months, in February of 1891, at the age of forty-one, she was dead of pneumonia. Within the next few years the book was translated into French, German, Dutch, Danish, Polish, Czech and Japanese. Two English translations (riddled with errors of the type made notorious in the parodies of Vladimir Nabokov) appeared in 1895: one in London, the other in New York.

For the contemporary reader Kovalevskaya's memoir still retains its original freshness as a personal document. At the same time it vividly recreates a segment of the social history of the period in its portrait of a wealthy landowning gentry family, struggling to preserve its stability against the erosions of an era which witnessed the emancipation of serfs from their masters, the rebellion of children against their fathers, and the rise of radical political groups armed with limitless faith in the power of science to bring about a just social order.

The English version presented here was translated from the 1974 Russian edition of *Vospominaniya detstva* contained in S. V. Kovalevskaya, *Vospominaniya i povesti,*

Nauka, Moscow, 1974 (the most complete collection to date of Kovalevskaya's fictional and critical writing), with a few corrections of minor errors discovered in comparing the 1974 text with the editions of 1951 and 1961 edited by S. Ya. Shtraikh. It includes the first translation into English of the chapter "Palibino" from Kovalevskaya's own text. The early Russian editions did not include this chapter, and the Shtraikh editions (published before Kovalevskaya's original manuscript of the chapter was discovered in the archives of the USSR Academy of Sciences) had utilized a translation from Swedish into Russian done by Kovalevskaya's daughter, Sofya Vladimirovna.

Certain other revisions should be mentioned as well. In the first Russian editions some of the chapters were left untitled, and Shtraikh supplied his own titles for these. The translator has substituted titles more appropriate to the content of the chapters. The chapter called "Palibino," which had been inserted by Shtraikh as Chapter Seven, now stands as Chapter Four, where it seems to fit best chronologically. The present translation also names the family's Polish tutor, who remained nameless in Kovalevskaya's manuscript for reasons described in the notes. In a few instances, sentences or paragraphs which appeared in the Swedish edition, but not in the Russian, have been included in the translation. These insertions are indicated wherever they occur. Finally, the book's title itself was changed to *A Russian Childhood*, in the hope of evoking some more concrete association in the minds of English-speaking readers than is called forth by the name of Kovalevskaya, which by now is almost unknown outside of mathematical circles (where she has been cited as "the most important woman mathematician prior to the 20th century").

Kovalevskaya's memoir comes to a close with the Dostoevsky episode, when she was fifteen years old and only just beginning to have some intimation of the possibilities of adult life opening up before her. In order to fill out the picture, therefore, supplementary material has been provided. This includes Kovalevskaya's own *Autobiographical Sketch,* an account of the highlights of her scientific career published posthumously in *Russkaya starina* and prepared for publication by her brother Fyodor.

Academician P. Y. Polubarinova-Kochina, of the Institute for Problems in Mechanics of the USSR Academy of Sciences, has contributed an article explaining the significance of Kovalevskaya's mathematics.

Finally, the translator has supplied a biographical introduction which attempts to set Kovalevskaya's memoir into a broader social and historical context. It first appeared, in briefer and somewhat different form, in *Russian Literature Triquarterly* ("Sofya Kovalevskaya: Growing Up in the Sixties," No. 9, Spring, 1974, 276–302).

Many people have contributed valuable help and suggestions in the preparation of this book; none of them bears responsibility for its deficiencies. I would like first of all to acknowledge my indebtedness to those scholars who contributed articles and commentary to the various Soviet editions of *Vospominaniya detstva*: most particularly to S. Ya. Shtraikh, through whose work I first became acquainted with the complex and fascinating interaction of the Korvin-Krukovsky and Kovalevsky families with the larger society they inhabited.

My gratitude is due to Dr. Neal Koblitz of the Mathematics Department of Harvard University and now a fellow-researcher in Moscow, for the graciousness with which he undertook to translate Academician Kochina's article into English. Special thanks go to Joan Thatcher (Zhannochka Tetcher) and to Sharon Miles, for aid and comfort when it was most needed. As ever, I am in debt to Rose Raskin of Columbia University: valued reader, stern critic and priceless friend. To Michael Stillman, who is at the center of my life, there is no adequate way to acknowledge or measure my gratitude.

BEATRICE STILLMAN

Moscow
May, 1978

Introduction

In mid-July of the year 1868 Vladimir Kovalevsky, a high-minded young Petersburg publisher of Darwin, Huxley and other European scientists, sent a rather peculiar letter to his fiancé, Sofya Korvin-Krukovskaya, at her family's landed estate in the province of Vitebsk.

"My first business in Petersburg, of course," said Kovalevsky, "will be to make an inspection and selection of the most suitable materials for the preparation of preserves, as per your commission. We shall see how this new product will succeed."[1]

Was Kovalevsky planning to abandon his moral principles and undertake a commercial venture with his gently-born betrothed? Not at all. The letter was a coded message designed to throw Sofya's father off the track, in the event that he should exercise his paternal right to read his eighteen-year-old daughter's mail. "Preserves" was the agreed-upon code word for husbands: in this instance, fictitious husbands, a desperately needed commodity for young women burning to liberate themselves from parental supervision in order to leave Russia and train for professional careers.

Vladimir's commission consisted of nothing less than locating a man of good family, a partisan of equal rights for women, who was prepared to sacrifice his own chances for a happy personal life[2] by entering into a fictitious marriage with Sofya's elder sister Anyuta and giving her his name, his protection and a foreign passport. This was to be Anyuta's ticket to freedom from General Korvin-Krukovsky's patriarchal household into a society more receptive to her aspiration of earning her living by writing.

Sofya herself had embarked on her relationship with Kovalevsky by the same route. Endowed with an extraordinary creative intelligence, and born into a society which closed the doors of its universities to women, she passionately wished to pursue a career as a doctor or a chemist so that she could "be of use." The only way to accomplish this was to enter a Swiss or German university, as a few other Russian girls had already done, then return to Russia with her degree and begin to practice a profession. And the only means of wresting her parents' permission to go abroad, it seemed to her, was through a fictitious marriage.

Like-minded friends in Petersburg (where the *fiktivny brak* had reached something of a vogue in radical circles) acquainted Sofya with Kovalevsky, an ardent disciple of Darwin and would-be geologist,[3] who was delighted to lend his person to the cause of Sofya's liberation. A way was found to insinuate him into a drawing room socially acceptable to General Korvin-Krukovsky. The courtship progressed, largely through letters in the vein of the one quoted above; the resistant but ultimately helpless General finally consented to the marriage, and the wedding took place in September, 1868.

And so the emancipation of Sofya Krukovskaya was accomplished through the literal application of sexual politics. The term is used here not in its current connotation of an instrument employed by men for the oppression of women, but in precisely the opposite sense: a means of utilizing existing Russian law for the purpose of helping women to become free.

In her aversion to all forms of established authority, whether vested in the monarchy, the landed gentry or one's own parents; her zeal for social reform and desire to render practical service in that struggle; her advocacy of individual rights while denying the validity of personal emotions; her passionate concern with the "women question," her infatuation with the natural sciences, and in particular her sanguine trust in science itself as the truth that would make men free, Sofya was merely behaving like a conformist, to the degree that these articles of faith were part of the repertoire of the "girl of the sixties." What was not at all representative was the palpability of her creative endowments: she was a gifted writer and an authentic mathematical prodigy.

Hardly had the wedding bells died down and the euphoria of her new freedom subsided, however, when Sofya became aware that her problems had only just begun—were, indeed, so vast in dimension that their outlines could not even be delineated. She had yet to learn what it meant for a woman, in the second half of the nineteenth century, to carry on a professional career in the masculine outside world while retaining a sense of femininity and continuing to function within the traditionally prescribed feminine realms of marriage and family.

In the process of her learning she so intensely experienced external obstacles and interior pangs of guilt and loss, struggled so hard to integrate the disparate areas of her life, achieved such spectacular successes while retaining a sense of constant failure, that her life might serve as an exemplar for a feminist history. She lived near the center of the social and ideological ferment which agitated her time. By virtue of her innate gifts, breadth of interests and extraordinary network of familial connections, friendships and intimate relationships, her life meshed in one way or another with some of the leading figures of her era in literature, science and politics: Dostoevsky, Strindberg, George Eliot, Helmholtz, Darwin, Mechnikov, Mendeleyev, Chernyshevsky, the leaders of the Paris Commune of 1871. In her chosen field of mathematics she was on close terms with the foremost mathematicians of her day: Weierstrass, Hermite, Poincaré, Picard, Chebyshev, Cantor, Kronecker, Mittag-Leffler and many others.

A bare enumeration of the achievements Kovalevskaya crammed into her life (which came to an abrupt end at the age of forty-one) is in order. Without ever being permitted to matriculate at a university, she earned a doctorate in mathematics entirely on the strength of her original contributions to the field. She was one of the first women in history to hold a university professorship (at the University of Stockholm, from 1884 to the end of her life). She was awarded the Prix Bordin of the French Academy of Sciences for a substantial contribution to an unsolved problem in mathematical physics. In succeeding years she became an editor of *Acta Mathematica,* was honored by the Swedish Academy of Sciences, and—perhaps the highest accolade of all in view of the formidable opposition mounted against her—she was the first woman to become a corresponding member of the arch-conservative Russian Imperial Academy of Sciences. In order to accomplish this, the leading Russian mathematicians of the day had to move heaven and earth to have the Academy's charter amended. When, after her election, Kovalevskaya tried to attend a session, she was denied admittance on the grounds that it was not in the tradition of the Academy to have women present at its meetings.

In the field of literature, in which she began to work seriously only during the last five years of her life, she produced the memoir of her childhood and adolescence presented in these pages; a novel, *A Nihilist Girl*; two plays, a personal reminiscence of George Eliot and a critical article on M. E. Saltykov-Shchedrin; a small body of verse and a collection of short stories, sketches and journalism. At the time of her death she had a number of major projects, including a novel about Chernyshevsky, in various stages of completion or still in the planning phase.

Despite the unfinished, topical or fragmentary nature of much of her work, *A Russian Childhood* alone clearly indicates her promise as a writer, had she been granted the normal span of years in which to develop it.

Sofya Vasilievna Korvin-Krukovskaya was born in Moscow on January 15, 1850. She was the middle child of General Vasily Vasilievich Korvin-Krukovsky, an artillery officer who in 1843 had inherited two sizeable estates, Palibino and Moshino, comprising together about 3500 desyatinas (9400 acres) in the then province of Vitebsk. Shortly after receiving his inheritance he married a much younger woman, the beautiful, gay and musically accomplished Elizaveta Fyodorovna, granddaughter of the astronomer F. I. Shubert. There were two other children: the strikingly beautiful Anyuta, seven years older than Sofya, and a younger boy, Fedya.

In 1858 Vasily Vasilievich retired from army service and took his family to Palibino, where rumors of the impending emancipation of the serfs impelled him to occupy himself seriously with the management of his estates. These included herds of cattle and sheep, an active dairy farm, a vodka distillery, flower and vegetable gardens, forests abounding in game, and lakes stocked with fish which were taken for sale to Vitebsk and Dvinsk.

The new problems, added to the Polish Uprising of 1863—the repercussions of which were strongly felt in a province at the very border of Lithuania and where many of the neighboring landowners were Poles—gave the family's life in Palibino a darker coloration than it had previously had. Still, the Korvin-Krukovskys lived comfortably and well in their manor house with its double-storied wings flanking it on both sides, triple-storied tower, ballroom, home theater, greenhouse, guest rooms and servants' quarters. As leaders of the local gentry (the General held the position of Marshall of the Nobility for the Nevel district and later for the province) they entertained lavishly with dinners and balls, home-produced plays and *tableaux vivants.* The children were brought up by their nanny until their education was taken over by governesses and tutors.

The first of these was a French mademoiselle. She was replaced by an English spinster, a Miss Margaret Smith, whose purview included English, French, music and deportment. There was also a resident tutor, Y. I. Malevich, a dedicated pedagogue who taught the children history, geography, arithmetic and geometry, Russian language and literature.

At thirteen Sofya began to exhibit an aptitude and avidity for algebra which alarmed her father, who had a horror of learned women, and it was decided to put a stop to any further mathematics studies. But she managed to borrow Malevich's copy of Bourdon's *Algebra* and kept it out of the governess's sight, reading it secretly at night after the household was asleep. An even more alarming episode occurred about a year later. A neighbor, Professor N. N. Tyrtov, came to visit and presented the family with a copy of his new physics textbook. Sofya took it and began reading, but when she reached the section on optics she was stopped by the presence of trigonometric formulas and terms she had never heard of.

She went to Malevich and asked him what a sine was. Since he was a systematic man and since trigonometry did not enter into his curriculum, he told her he didn't know. She thereupon returned to the book and tried to explain it to herself in her own way, using concepts with which she was familiar. For the enigmatic sine, she substituted a chord—a quantity closely approximating a sine in the case of small angles. And since all of Tyrtov's formulas employed only small angles, her method worked.

On Professor Tyrtov's next visit Sofya tried to draw him into a discussion of his book, but he responded reluctantly and finally said straight out that he found it hard to believe she had really read and understood it. At this her adolescent amour propre was wounded, and she insisted that it had presented no special difficulties for her. "Now you're bragging, aren't you?" asked Tyrtov. But when Sofya explained the way she had gone about trying to make sense of the material, Tyrtov was dumbfounded. In a state of high excitement he went to the General and said that Sofya must be given serious training in mathematics, pointing out that in her independent working out of the concept of "sine" she had used the same method by which it had developed historically; and he even went so far as to compare her with Pascal.

Vasily Vasilievich hesitated over the decision for a long time, and finally, several years later, he permitted her to take private lessons with Professor A. N. Strannolyubsky in analytic geometry, differential and integral calculus during the family's annual sojourn in Petersburg.

Perhaps the deepest and most disturbing influence on Sofya's development during her late childhood and early adolescence was the upheaval in the Krukovsky household concomitant with her sister Anyuta's stormy maturation into young womanhood. When Anyuta was about twenty she became deeply engaged with radical ideas, which had belatedly penetrated to Palibino from the distant capital. The chief transmitter of these new views was the local *popovich* (son of the priest), home on vacation from the university, who scandalized the countryside with asseverations of man's origin from apes and with the eminent Professor Sechenov's[4] proof that man has no soul, only reflexes.

It was her association with this young man which precipitated Anyuta's first act of open rebellion against her father. It also activated dreams of leaving home to study in Petersburg and even vaguer fantasies of joining Vasily Sleptsov's notorious commune on Znamenskaya Street, where, it was rumored, well-born young people of both sexes were living together, caring for themselves without servants and learning to perform manual labor in a state of total communism.

Frustrated in these aspirations by her father's refusal to allow her to live in Petersburg, Anyuta took on another occupation: writing stories in secret under a male pseudonym, with the fourteen-year-old Sofya as her eager confidante. Her work evidently showed some talent, for two of her novellas, "The Dream" and "Mikhail," were accepted for publication by Dostoevsky in his journal *The Epoch* for 1864.

By 1868 Anyuta, now twenty-five years old and still "incarcerated" (the word is the sisters' own and appears more than once in their correspondence) at Palibino except for the annual visit to the Shubert aunts in Petersburg, had somehow managed to strike up a personal acquaintance with a radical group, staunch advocates of higher education for women, through whom she got wind of the possibility of a fictitious marriage as the escape route from legal subjection to her father. Prominent in the circle were Nadezhda Suslova (sister of the celebrated Polina), who was to become the first woman physician in Russia, Dr. P. I. Bokov, and his wife Maria. In the course of the lectures she attended at the Medical-Surgical Institute Maria had fallen in love with her physiology professor, I. M. Sechenov (that same notorious Sechenov who believed in reflexes and not souls); and the result was a *menage à trois* so friendly, cooperative and durable that it became the model for Chernyshevsky's novel *What Is To Be Done?*[5]

It was Maria Bokova who, in the spring of 1868, introduced Anyuta and her best friend Zhanna Evreinova[6] (who was later to become the first woman lawyer in Russia) to Vladimir Kovalevsky as a possible fictitious husband. There were several surreptitious meetings during which Anyuta and Zhanna explained to Vladimir that it was immaterial which of them he decided to marry, since the other would then be able to come along with them to Europe, properly chaperoned by a married couple.

On one of these occasions Anyuta brought the eighteen-year-old Sofya with her; and after that meeting Vladimir overturned all the previous negotiations by insisting that he would marry Sofya and no one else.

It is evident that Kovalevsky had violated a canon of radical ethics by allowing personal feelings to intrude into his decision. Whether or not he was influenced by Sofya's blend of unnerving intelligence and fledgling vulnerability (he immediately started referring to her as "sparrow")—as contrasted, perhaps, with Anyuta's poised beauty and Zhanna's rather formidable purposefulness—it is clear that he was smitten with Sofya.

The convinced materialist writes:

> Meeting you makes me believe in the affinity of souls, so swiftly . . . and genuinely did the two of us come together and, on my part at least, become friends. . . . Now I cannot keep from picturing much that is joyous and good in our common future. Indeed, employing the coolest possible judgment, without childish enthusiasm, one can state almost positively that Sofya Vasilievna will become a splendid doctor or scholar in some branch of the natural sciences; it is further quite possible that Anna Vasilievna [Korvin-Krukovskaya] will be a gifted writer, that Nadezhda Prokofievna [Suslova] and Maria Alexandrovna [Bokova-Sechenova] will be excellent physicians, that Iv. M. Sechenov will always remain (for some) or become (for others) our mutual and dearest friend, that I, your obedient servant, will

> expend all my powers on the flowering of our union; and you can picture for yourself what brilliant possibilities of happiness, how much good, sensible work are in our future.
>
> So you see that I fuse not only my own, but also the interests of the above-mentioned persons with yours, and I dare say that it is precisely such a fusion which will secure a good future for us. Therefore you should look on me now, not as a man doing you a favor, but as a comrade striving jointly with you toward a single goal; that is, I am exactly as necessary to you as you are to me; therefore, make use of me accordingly, and entrust to me whatever you may take into your head without fear of burdening me; I shall work just as much for you as for myself (VIP, 486).

For Sofya the world had been stood on end. At one stroke she was presented with personal freedom, an attentive young husband who seemed to be on intimate terms with the entire science faculty of the University of Petersburg, and the prospect of pursuing her higher education without hindrance or limit. Moreover, it now devolved upon her to find a husband for Anyuta; and this reversal of their respective positions, with all the implications flowing from it, was difficult for her to handle. Throughout her life she had acknowledged her sister to be her unquestioned superior in beauty, talent and lovableness; and if her adoration concealed some jealousy, she had permitted it to come to the surface on only one occasion: during the period three years earlier when Anyuta was being courted by Dostoevsky.

On top of all that, it was Anyuta who had shown the courage to fight for independence, but Sofya to whom it had been presented as a gift—a paradox indeed, and a new source of guilt. Her letters from Petersburg in the fall of 1868, just after her marriage, to her "dear, darling and priceless" sister, still shut away in Palibino, pour out in a flood of extravagant, repetitious expressions of love mingled with guilty admissions of undeserved happiness and high-pitched descriptions of her new life:

> Sechenov's lectures begin tomorrow; and so my real life begins at 9 A.M. . . . [Vladimir Onufrievich] and friends will solemnly escort me by way of the back stairs so that there is hope of hiding from the administration and from curious stares (VIP, 225).

> I forgot to tell you that Mechnikov promised to admit me to his lectures and get permission for me to attend the physics lectures. . . . I'm studying physiology and particularly anatomy; we got a skeleton from Pyotr Ivanovich [Bokov] and brother is poking it at this moment. . . . Sometimes it's horribly painful for me to be without you (VIP, 229).

> You are necessary to me, Anyuta, more necessary, believe me, than you ever were in my life . . . I *must* not be happy without you! At times a strange anguish comes over me and I feel ashamed that everything is coming to me so easily and without any struggle . . . and it makes me angry that I have given myself so totally to the empty and frivolous side of my happiness. . . . Mechnikov came to dinner and we had a long talk . . . in general, we were very gay (VIP, 232).

At the end of September Sofya met Vladimir's elder brother Alexander for the first time. He was then twenty-eight years old and teaching at the University of Kazan; he had already done pioneering work in comparative embryology and was to go on to a distinguished career. Vladimir had unsuccessfully tried to entice him away from field work at a marine biological station to attend the wedding at Palibino; now, he dropped in unexpectedly to the newlyweds' apartment. Of Alexander's visit Sofya writes to Anyuta:

> He's not at all as I imagined him—quite microscopic and, in my opinion, not a bit like our brother [i.e., Vladimir]; he's handsomer, but his face isn't nearly as nice and sweet . . .

> although, generally speaking, I think Alexander Onufrievich is a very nice man. He went on and on to Brother about the necessity of close attachment and love between children and their parents, but that's forgiveable in the father of a family. As for the rest, he's a strong Nihilist and he advised me, in case they exclude me from Ivan Mikhailovich [Sechenov's] lectures, to dress myself in male clothing. (VIP, 233)

At this period of the marriage the relationship between Vladimir and Sofya seems to have resembled a preadolescent's fantasy. He lavished on her an inexhaustible fund of tenderness, pride and solicitude, all that emotional sustenance of which she had been starved in childhood and for which she grew increasingly insatiable as the years went on; and she was inseparable from him and addressed him constantly as "brother." "You can't believe what a nice; wonderful brother he is. I feel as though he has always been my brother" (234). She jokes about the legal impossiblity of sharing him with Anyuta: "Brother and I keep lamenting the fact that he's not a Mohammedan. That would have been marvelous" (226).

Meanwhile their plans were taking shape to cross the border to Germany and freedom in April of 1869; but no suitable marriage candidate could be found either for Anyuta or for Zhanna. Mechnikov, whose eligibility had been thoroughly discussed and approved, disappointed them by selecting a bride of his own; and Maria Bokova, that feminist militant on whom the sisters had been counting heavily, refused for some inexplicable reason to consider the marriage of her lover Sechenov to Anyuta, even though he was legally a bachelor. Once again private feelings had overridden rational considerations. In the end Sofya and Vladimir had to leave for Germany taking Anyuta with them, still a spinster. Zhanna, whose father had declared he would rather see her in her grave than in a university, remained in Russia for some months longer. Finally in mid-November she ran away from home without a passport and crossed the border on foot, by night, through a swamp, under fire from the Russian border patrol.

Shortly after her arrival in Germany Anyuta took off for Paris in search of revolutionary action, handing over to Sofya the task of forwarding her mail to their parents in Russia, so as to maintain the pretense that she was studying in Heidelberg. The secret leaked out, however, and General Korvin-Krukovsky stopped Anyuta's allowance. She thereupon found a job as a typesetter with a Paris printing press, a step which brought her into immediate contact with the working-class people and the revolutionary circles she had been seeking.

Sofya, alone in Heidelberg awaiting Vladimir's arrival from Vienna, then discovered with shock that university matriculation was not open to women. After enormous effort and a series of rejections that would have crushed anyone less persistent, she was able to wring from the university authorities permission to attend lectures on an unofficial basis. Soon, together with her new friend, Yulya Lermontova (a cousin of Zhanna who was eager to study chemistry) she was studying with the foremost scientists of Europe—chemistry with Bunsen, physics with Helmholtz.

Yulya, whose memoirs provide a valuable record of Sofya's student years, writes: "Sofya immediately attracted the attention of her teachers with her uncommon mathematical ability. Professor Königsberger, the eminent chemist Kirchhoff . . . and all the other professors were ecstatic over their gifted student and spoke about her as an extraordinary phenomenon. Talk of the amazing Russian woman student spread through the little town, so that people would often stop in the street to stare at her" (VIP, 381–83).

For a period of several months the trio—Sofya, Vladimir and Yulya—shared an apartment and led a comradely, industrious and at the same time lighthearted existence that Yulya later described as "unalloyed delight" (VIP, 377). She refers to Sofya's relationship to her husband during this time as "poetic." "[He] loved her with a totally ideal love, without the slightest tinge of sensuality. She, it was evident, felt an identical tenderness toward him. Both were, as yet, obviously alien to that sickly, base passion which is usually given the name of love" (VIP, 382). This idyll came to an abrupt end with Zhanna's and Anyuta's arrival in Heidelberg in November of 1869. The two newcomers at once made clear their antagonism to Vladimir and their hostility toward what they considered the couple's "excessive intimacy." Their interference in the Kovalevskys' mode of life eventually created an atmosphere so rancorous that Vladimir moved—first to separate quarters and then out of Heidelberg altogether—to other German universities, where he explored various branches of geology in search of an approach which would help to substantiate Darwinian theory.

The community of women was not of long duration. Zhanna, unsuccessful in her attempt to gain admission to Heidelberg, moved on to Leipzig. Anyuta returned to Paris, where she soon found herself in the thick of revolutionary activity. She became a close friend of the Socialist writer André Leo, joined the newly founded Society of Struggle for the Rights of Women, and worked with Leo in publishing a newspaper, *Women's Rights.* At about the same time she fell in love with Victor Jaclard, a fiery revolutionary, friend of Blanqui and Bakunin and later of Karl Marx.

Anyuta fully shared Jaclard's views and his revolutionary activity. Just before the outbreak of the Franco-Prussian war in July of 1870, she fled with him to Geneva to escape his imprisonment in connection with massive government reprisals for the murder of a police agent. From Geneva she repeatedly wrote to her parents in Palibino, asking them to provide her with the necessary documents so that she and Jaclard could be married, but there was no reply. She joined the Russian section of the International, wrote stories for children and translated pamphlets by Marx. Jaclard gave language lessons and waited the opportunity to return to France.

On September 1, 1870, Anyuta wrote to Sofya:

> The latest news of Napoleon's capture has us worked up even more than before, and we have decided to leave for Paris . . . Of course, the decision is fraught with danger. The Republic is still so far away that the new government, whatever it turns out to be, will be equally hostile to the revolutionaries. . . . But there is no other way. If a man wants his beliefs and actions to be accepted as real, he has to risk himself. And even if I had the power to hold Jaclard back, I would not use it under any circumstances.[7]

Anyuta's estimate of the remoteness of the Republic was in error. It was proclaimed simultaneously in Paris and Lyons on September 4, 1870—the very day of the Jaclards' arrival in Lyons. The war progressed with disastrous rapidity for France. Within two months Strasbourg and Metz fell to Germany. Members of the government fled the capital (Gambetta by balloon). By the end of January France had surrendered all its forts, and Paris was surrounded. Adolphe Thiers, appointed Premier, withdrew the government and troops to Versailles, and a starving city was left to defend itself against the besieging Prussian army.

Jaclard, who had close connections with the workers' movement in Lyons, had been elected a delegate of the "Red Lyons Republic" almost immediately upon his arrival in France and was sent to Paris to serve as liaison with the National Guard. Once there, he was quickly drawn into the organization of the twenty *arrondissements* of the city for defense against the siege, was elected to the Constituent Assembly and appointed commander of the National Guard troops for the 18th (Montmartre) *arrondissement,* the center of working-class strength.

After his surrender of France to Prussia in March Thiers, who considered the crushing of the National Guard his first objective in the consolidation of the Republic, was free to turn the full force of his troops and cannon against Paris. But the city which had withstood the German siege resisted Thiers as well. On March 18, 1871, the National Guard proclaimed the Commune of Paris and defended the city for 72 days (under bombardment by Thiers' cannon from April 1)—up to the 28th of May.

During the ten weeks of the Commune's existence Anyuta was no less active than Jaclard. As secretary of the Committee of Vigilance of the 18th *arrondissement* in charge of hospitals, schools and prisons, she nursed the wounded, dismissed the nuns (supporters of the monarchy) from their hospital and teaching posts, rewrote curricula for girls' schools, issued a proclamation outlawing prostitution, and (with André Leo) founded a newspaper, *La Sociale,* which came out daily between March 31 and May 17.

At about the same time that Anyuta was living in Geneva with Jaclard, Sofya had come to a momentous decision for herself: that her true vocation was mathematics and that there was one mathematician above all others in the world she wanted to study with—Professor Karl Weierstrass, of the University of Berlin. Armed with recommendations from Königsberger and other Heidelberg professors, she went to see Weierstrass in August, 1870.

She quickly discovered that if university affiliation had been difficult for women at Heidelberg, at Berlin it was impossible. The university's regulations barred women without exception, including even unofficial and occasional attendance at lectures, and this despite the fact that the war had decimated the students' ranks to the point where classroom attendance was very thin. But Weierstrass was so deeply impressed with Sofya that, notwithstanding a strong prejudice against higher education for women, he agreed to teach her privately. Within a few months he had come to regard her as the most brilliant and promising of all his students, and was sharing with her not only his university lectures but also his ideas and his unpublished work.

But the work with Weierstrass was interrupted by Sofya's terrible anxiety for her sister, the most beloved person in her life. In April of 1871 she and Vladimir, under mind-boggling conditions, returned to Paris, somehow managing to infiltrate the German troops still quartered in the vicinity. Partly on foot and partly by boat down the Seine, they reached the besieged city and entered it on April 5. Between then and May 12 (approximately half the lifetime of the Commune itself) they lived with the Jaclards, where Sofya worked side by side with Anyuta in the hospitals and Vladimir walked the city's devastated streets and even found time to study paleontological collections in its museums. Later in her life Sofya often spoke of writing a memoir of that period but was never able to realize her project.

Believing that the Commune would be able to hold out for a few months longer, or that Thiers would work out some kind of armistice with the National Guard rather than destroy the city, the Kovalevskys decided to return to Berlin and their scientific work. Again the prediction was mistaken. On May 22 the Versailles troops entered Paris at Auteil, and by the last days of the month the Commune had been annihilated in a bloodbath in which over 20,000 Parisians were executed without respect to age or sex.

A letter arrived from Anyuta informing them that Jaclard had been arrested and was in prison facing a death sentence. The Kovalevskys returned to Paris at once. In a burst of the misdirected but heroic altruism which characterized and eventually destroyed his life, Vladimir worked out the following plan, which he communicated to his brother Alexander in a letter of June 13:

> In such a situation Anyuta, of course, could not be of any use, and so we got her out of Paris immediately and stayed behind ourselves to see what we could do for him. Yesterday I went to Versailles, where he's in jail, but I couldn't get to see him. . . . There isn't the slightest doubt that he'll be deported. Where to?—We don't know. Probably to New Caledonia. Anyuta of course will follow him there. But since he'll be sent together with the other prisoners on a transport ship around the Cape of Good Hope, Anyuta will have to travel alone, which I think will not be possible. Sofya is dying to go with her. This I consider an absurd idea, since it would keep her from finishing her examinations, which could come about in the next six or eight months. So obviously . . . the force of circumstances dictates that I am the one who should accompany Anyuta through Suez, Ceylon and Melbourne. Besides, since I'm a free man, it's I who should settle in New Caledonia with them, and Sofya can come to us after she passes her examinations. It's the only possible way. Sofya and Anyuta have become my own kin, so that I cannot separate from them.[8]

The "force of circumstances" dictated otherwise. On being informed of their daughter's plight General Korvin-Krukovsky, then seventy years old, made the trip from Palibino to Paris together with his wife. The General was an old acquaintance of Thiers; what transpired between them is unknown. Several contradictory versions are given for the manner of Jaclard's escape from prison in Chantiers, all of them highly implausible. The fact remains that he did escape and reached Switzerland, using the passport loaned him by Vladimir Kovalevsky. The General at last gave his consent to the marriage and departed with his wife for

Palibino, leaving the legally wedded Jaclards in Geneva.[9] Vladimir returned to his paleontological studies, and Sofya—to Weierstrass in Berlin, where Yulya Lermontova was waiting for her.

Yulya had also managed to make a private arrangement to continue with her chemistry studies, and the two women settled down in miserable quarters to an ascetic, monotonous and isolated existence which seems to have consisted almost exclusively of grinding, unrelieved labor. It was then that Sofya, according to Yulya, began to exhibit that single-minded fixity of purpose of which there had been hints even during her childhood years, but which now burst forth with full vigor. Her attitude toward work assumed the proportions of an *idée fixe;* she had an unnerving capacity to sit at her desk for limitless hours of intense concentration, from which she would arise exhausted and depressed. Deprived of Vladimir's leavening influence, she reached the point of renouncing every distraction and was reluctant even to leave the house.

By 1874 Sofya, with characteristic prodigality and compulsiveness, had produced not one but three original works, of which Weierstrass said that he would have considered any one entirely acceptable as a doctoral dissertation. The problem was where and how to present them. Berlin was hopeless. Weierstrass favored the University of Göttingen, where a precedent existed for awarding doctoral degrees to foreigners in absentia. Göttingen, after raising fusillades of objections, finally capitulated. In July of 1874 it took the unprecedented step of awarding to Sofya—without examination or defense, entirely on the strength of the three works she had presented—the degree of doctor of philosophy in mathematics, *summa cum laude.*

Yulya received her doctorate in chemistry from Göttingen in the same year. Vladimir had earned his own doctorate two years earlier from the University of Jena with a brilliant dissertation on the paleontology of ungulates and had produced several pioneering works whose importance was acknowledged by Darwin. During the six years of their marriage the couple's relations had grown increasingly troubled and complex. Their early lighthearted devotion had eroded under the pressures of financial strain, geographical separation, and the growing resentment of each of them at running a poor second to the work of the other. And then there was, of course, sex. A letter from Vladimir to his brother Alexander written apparently some time in 1871 throws a good deal of light on this subject.

> I love Sofya deeply although I wouldn't say I was what they call "in love" with her. . . . During our life together I might have made myself her husband, of course, if I had wanted that very much, but I was always very much afraid of it for many reasons. . . . It wouldn't have been right, somehow, coming together in the way we did and getting married out of necessity, suddenly to change to a genuine marriage; it would have been like taking her by cunning, and that wouldn't have been pleasant for me.

He goes on to speak of Sofya's fear that pregnancy would put the finish to her work, and of his own unsuitability as a husband and father. This very logical reasoning was superseded by its opposite at the end of the German interlude, six years after their wedding, when the Kovalevskys decided to "transform the fictitious marriage into a real one" (as Russian literary historians delicately put it). The decision was based by this time less on desire than on honor, feelings of obligation to Sofya's parents, the importance of putting an end to "lying relationships" and similar abstract considerations, and it was a bad one.

Bearing all this heavy intellectual and emotional baggage, the expatriates returned home to Russia in the summer of 1874, that notorious "mad summer" when Russian students by the hundreds made their futile pilgrimage "to the people," to be met by suspicion, antagonism and Tsarist police.[10]

The normalization of the Kovalevskys' marital relations marked at the same time the most contradictory period of Sofya's career. Having violently dislocated her personal life to get a higher education and then given up five more years of it in monastic dedication to the mastery of her discipline—in the course of which she achieved a success so spectacular and so tradition-shattering that it made her something of a celebrity—she packed her two still unpublished dissertations into a box where they proceeded to gather dust for the next six years; and she tried to model herself into a conventional Petersburg matron.

Why she did this remains an intriguing question for a biographer. Her first memoirists tend to ascribe the six-year hiatus to "overexhaustion." An early biographer, E. F. Litvinova (another woman mathematician who was inspired by Sofya's example back in 1868 to pursue her own career) notes more perceptively that it was characteristic of Kovalevskaya to set herself unattainable goals, but not to take pleasure in their fruition (*Sofya Kovalevskaya,* St. Petersburg, 1893). Contemporary feminists would probably speak of "fear of success." A Soviet biographer (avoiding confrontation with the irrational elements in her subject's personality) emphasizes the near-impossibility of a woman's being allowed to teach in Russia at that time. This last consideration is important and must not be discounted; but the fact remains that Kovalevskaya not only did not teach: neither did she go on with her own research, nor did she reply to the letters of Weierstrass, who regarded her as a spiritual daughter and who was appalled at this waste of her gifts.

The first winter in Petersburg, with its release from the terrible strains of her grindingly tense, concentrated Berlin existence, marked a great opening out of her personality. She uncovered in herself a thirst for friendship, talk, gaiety and diversions of all sorts, and at the same time a talent for them that invariably made her the center of the literary and intellectual circles in which the Kovalevskys traveled.

Finding work was another matter. On discovering that the best job that could be expected was teaching arithmetic to the elementary classes of a school for girls, she remarked bitterly, "I was, unfortunately, weak in the multiplication table."[11]

Nor was Vladimir successful in obtaining the hoped-for university professorship, despite his brilliant European degree. Through a group of wealthy acquantances the Kovalevskys kept hearing stories of fabulous profits to be made in buying and selling houses. They began to construct a fantasy of getting rich quickly so that they could use their accumulated means to work independently in science at some future date. Using the inheritance that had come to Sofya on the death of her father, Vladimir entered the construction business. His capital was far exceeded by the grandiosity of his plans, and he was caught up in a spiral of rising debts and angry creditors, of mortgaging and re-mortgaging his still unbuilt structures.

During this altogether strange period in their lives, Sofya worked in two areas more consonant with her interests and her style. One of these was a two-year stint as theater reviewer and science reporter on A. S. Suvorin's newspaper, *Novoe vremya* (New Times)[12], during which she wrote on the latest achievements in science and technology: the flying machine, the typewriting machine, the telephone, solar heaters, the newest researches at the Pasteur Institute. She also attended the celebrated trial of the "One Hundred Ninety-three" (mainly of students who had taken part in the movement "to the people" during the summer of 1874), and collected material which she later used in her novel *A Nihilist Girl.* The second activity was her participation in 1878 in organizing the "Higher Courses for Women" envisaged by its founders as the nucleus of a women's university.

Both of the Kovalevskys were active in administrative and fundraising work for the school. With his characteristic over-generous zeal, Vladimir presented it with several copies of each edition in his publishing catalogue; there was even some talk of his constructing a new building to house it.

But this new work, too, was in the end to produce more bitterness for Sofya than gratification. After she had publicly announced her readiness to teach without salary, she was passed over as a member of the school's faculty. The result was a further blow to her self-esteem and the closure of an exit from a style of life she was now beginning to refer to as "the soft slime of bourgeois existence."[13]

In October of 1878 a daughter, Sofya Vladimirovna, was born and nicknamed Fufa by her doting parents, and for a time all Kovalevskaya's intense energy was focused on the rearing of the infant. It was not sufficient for her: she was too much of an individual and too committed to her work to content herself with an existence preoccupied exclusively with domesticity.

Vladimir's business affairs, meanwhile, were worsening drastically. Sofya was impatient and angry with what she considered his refusal to face facts, and as her dissatisfactions increased in number and depth she began to keep a notebook in which she caustically satirized them both: her husband for his comments on the incapacity of women to do creative work, and herself as "the erudite wife."[14] No more than a slight nudge was needed to bring their division out into the open.

That impetus was provided by the Sixth Congress of Natural Scientists held in Petersburg in January of 1880. P. L. Chebyshev, an eminent mathematician who was an old friend of the Kovalevskys, urged Sofya to give a paper at the Congress on some aspect of her work. Reluctant at first, she went back to her unpublished dissertation on Abelian integrals. In a single night she translated it into Russian and read her paper the next morning to enthusiastic approbation. Despite the six-year lapse since its writing, nothing superseding it had been published in the interim, and her findings were received with great interest by Russian mathematicians who had in the past displayed resistance to the direction taken by the German school. At one stroke she had reestablished herself as a mathematician to be taken seriously.

One of those who heard Sofya was Gösta Mittag-Leffler, a former student of Weierstrass and now Professor of Mathematics at the University of Helsingfors. He had come to the Congress charged with Weierstrass' commission to seek Sofya out and try to bring her back to serious work. Mittag-Leffler was deeply impressed with her scholarship and her personality; he spoke hopefully of finding her a place on the Helsingfors mathematics faculty. Full of her triumph, she was capable, two weeks after the close of the Congress, of being a composed witness to the sale of the last of the Kovalevskys' possessions at public auction, and the collapse of her husband's brief career as tycoon manqué.

But for Vladimir it was unmitigated disaster. He became, in the Russian idiom, "stiller than water and lower than grass." Failing to understand that his depression was a symptom of mental illness, she criticized him for weakness of character, while his brother Alexander rubbed salt into the wounds by asking rhetorically, "What might you not have achieved if all that energy had gone into paleontology?"

In Moscow, he began all over again to search frenetically and futilely for an academic position. Rejected everywhere, even for the post of curator of fossil collections in a small provincial museum, he was unable to resist his old fantasy of superior business acumen. He accepted a position with the petroleum firm of V. I. and L. I. Ragozin, authentic tycoons of that era of incipient industrial expansion in Russia. When, at the end of 1880, he was unexpectedly offered a post as docent in paleontology at Moscow University, he accepted it while staying on with the Ragozins: a juggling act that could not fail to end catastrophically.

It was the beginning of the end of the marriage. Sofya utilized her husband's prolonged absence on a business trip to make a long visit of her own to Weierstrass in Berlin, to discuss Mittag-Leffler's dazzling promise of a teaching post and to consult on ways of making up for six years of lost time.

The rest is a history of mutual absences and grievances. It was agreed that Sofya should live abroad "for a while." At a brief meeting in Anyuta's apartment in Paris, the separation was formally acknowledged. For the next two years Sofya—sometimes with the baby Fufa, more often leaving her in the care of Yulya Lermontova or Alexander Kovalevsky—lived an ascetic student life in dreary furnished rooms in Berlin or Paris, worked at a major new research project on the refraction of light in crystals, and waited for a teaching job. Mittag-Leffler's negotiations with Helsingfors on her behalf had fallen through, and he himself had transferred meanwhile to the newly founded university of Stockholm, where he continued to put forth such strenuous efforts for her that she became afraid he might undermine his own position.

It was an ascetic existence but no longer a solitary one. Thanks to the good offices of Weierstrass and Mittag-Leffler Sofya was put in touch with the most distinguished mathematicians of France, among them Charles Hermite and Henri Poincaré—and through them a circle of professional contacts was established. And there was a warm and stimulating social life as well in the colony of radical Russian emigrés now living in Paris, with Pyotr Lavrov and the Polish revolutionary Maria Mendelson at its center.

In the spring of 1883 Vladimir, despairing of his capacity to produce further creative work or even to retain his modest position on the Moscow University faculty, and faced with legal implication in the Ragozin stock scandal, killed himself by inhaling a bottle of chloroform.

When the news reached Sofya in Paris she shut herself inside her room and attempted to starve herself to death. She collapsed on the fifth day. When she was brought back to consciousness she reached for pencil and paper and immersed herself in mathematical computations. She was never thereafter to rid herself of guilt. Years later in discussing the event she still wept and blamed herself for having abandoned Vladimir during the greatest crisis of his life.

Shortly after Vladimir's suicide Weierstrass received a letter from Mittag-Leffler, informing him that the obstacles to hiring a woman at Stockholm had been overcome, and that Sofya was free to come at any time to begin her course of lectures. There was a condition attached to the offer. She could not be a member of the university faculty for the first year, during which she was expected to "demonstrate her competence"; she would be given the title of privat-docent, roughly equivalent in status to assistant professor, but without staff affiliation or salary, aside from private fees she might receive from her students.

Kovalevskaya's response to this more than modest offer was striking.

> . . . I consider myself in many respects extremely ill-prepared to carry out the duties of a lecturer. I have such deep doubts of myself that I am afraid that you, who have always

> shown me such good will, may be disenchanted when you realize how unsuited I am to my chosen work. I am deeply grateful to Stockholm University for opening its doors to me so graciously . . . but precisely for that reason I would not wish to come to you until I can consider myself completely worthy of the good opinion you have formed of me.[15]

In November of 1883, Kovalevskaya arrived in Sweden to begin her teaching career. Stockholm was itself a center of controversy between progressive and conservative factions. It was therefore inevitable that the appointment of "the queen of mathematics" (as she was ironically called in hostile academic circles) should be greeted by strong feelings on both sides. The writer August Strindberg wrote a newspaper article attacking Kovalevskaya on the grounds that "a female professor of mathematics is a pernicious and unpleasant phenomenon—even, one might say, a monstrosity; and her invitation to a country where there are so many male mathematicians far superior in learning to her can be explained only by the gallantry of the Swedes toward the female sex." (VIP, 363).

Her misgivings as to her competence to carry out her duties proved groundless. She was a natural teacher and had an enormous success, although she lectured in the abstruse and difficult area of her own specialty. Her lectures were given in German for the first year only; after that, in Swedish. In July of 1884, six months after her first lecture, she received her official appointment as Professor of Mathematics for a five-year period; she became an editor of the influential journal *Acta Mathematica* that same year, was given an additional appointment to the Chair of Mechanics in 1885, and a life appointment in mathematics in 1889.

The new circumstances of her life seemed to release

great untapped stores of physical energy and intellectual productivity. She began to study Swedish immediately upon her arrival, and within two weeks was able to hold a conversation of sorts in it; two months later she was consuming Swedish literature in great gulps. Her status as a celebrity brought her attentions from the intellectual elite and socially prominent, some of whom found themselves calling on her in her inevitable furnished room (not for another two years did she feel sufficiently secure to send for the dilapidated old Palibino furniture, set up an apartment of her own, and bring the eight-year-old Fufa to live with her for the first sustained period since the child's infancy). There was also a group of intimates drawn mainly from university circles. Her closest friend was Mittag-Leffler's sister Anne-Charlotte, a well-known novelist of advanced social views, deeply concerned with feminist themes. The two became inseparable friends, much addicted to self-observation and self-analysis. Anne-Charlotte said of Sofya that she tried to account to herself for every thought, feeling and act and to fit them into some psychological system; that she was always seeking connections. It was perhaps as a result of her relationship with Anne-Charlotte that Kovalevskaya began writing again, although she was careful to explain her motivation as a simple need for diversion from the tensions of mathematical research.

Her first piece, a personal reminiscence of George Eliot,[16] was occasioned by the publication of Eliot's letters in 1885. As a nineteen-year-old bride, Sofya had been taken by Vladimir to Eliot's Sunday at-homes in London. Her last meeting with Eliot took place just two weeks before the latter's death in 1880. On one of those visits Miss Eliot introduced her to a gentleman with the words, "Here is a female mathematician, the living refutation of your theory." She found herself excitedly defending the capacity of women for abstract thinking against the sceptical reactions of the stranger, who turned out, as she later learned to her chagrin, to be Herbert Spencer.

The next attempt was more serious and far more ambitious: a pair of parallel dramas titled *The Struggle for Happiness.*[17] The work was a collaboration between Sofya and Anne-Charlotte, calling themselves "the firm of Korvin-Leffler." The original idea was Sofya's, germinated out of the agonizing hours passed at Anyuta's bedside as she lay slowly dying of a chronic disease, while Sofya, observing her sister made unrecognizable by pain, dwelled obsessively on the way Anyuta's life had blasted her early promise.

"How it was" and "how it might have been": two plays using the same set of characters but two completely different plots, flowing out of the differences in the moral choices taken by the characters at critical points in their lives. The idea so excited her that she inflamed Anne-Charlotte's imagination with it as well. The pair embarked on a feverishly intense collaboration in which Sofya supplied the psychological materials and many of the incidents, the two of them together worked out the structure of their play, and the lines were written by Anne-Charlotte.

Sofya, whose mastery of Swedish was insufficient for her to take part in the actual writing, sublimated her frustration with embroidery, for which she conceived a passion at that time. While her fingers worked their design, scene after scene of the play would unroll in her mind. "The drama she could not write down herself with pen and ink she embroidered onto canvas with needles, silk and wool." (VIP, 438) When Mittag-Leffer came to visit and found her at her embroidery instead of at her computations, he would explode with exasperation.

The reason for his despair was the Prix Bordin competition, which he had decided his protégé must enter to secure her further career. This was an award offered by the French Academy of Sciences for substantive contributions to the solution of a long-standing problem in mechanical physics: the rotation of a solid body around a fixed point. The problem, whose elusiveness had earned it the epithet of "the mathematical water nymph," had intrigued and evaded mathematicians such as Euler, Lagrange and Poisson for years; a strict mathematical solution had been established only in certain isolated cases. The prize itself had been awarded ten times in the fifty years of its existence, and not at all in the previous three years.

Kovalevskaya's absorption in *The Struggle for Happiness,* however, preempted all other considerations. The play appeared in print in December, 1887, and received such acidly negative critical reviews that the theater withdrew its offer to put it on.[18]

In the autumn of 1887, Anyuta, after an operation described as "a complete success," died suddenly. Sofya kept repeating over and over again that the last thread had been cut connecting her with her own past. She complained of desperate loneliness, of being condemned to speak to her intimates in an alien language in which she could not express the nuances of her thinking and feeling, of being forced to go about wearing a mask on her face. That was her state of mind when Maxim Maximovich Kovalevsky[19] walked into her life in February of 1888.

He was a celebrity, but she had known many celebrities and was not easily impressed by them. He must have been a remarkable man, for she responded to him with all the pent-up feeling of which her nature was capable. He had been invited to deliver a series of lectures in sociology at Stockholm University, after being dismissed from his professorship in state law at Moscow because of suspected radicalism and excessive popularity with the students. He was an expert in English constitutional law, a long-time friend of Marx and Engels; his researches on the family and private property had been used by Engels in his *Origin of the Family, Private Property and the State.*[20]

Mittag-Leffler, desperate because Kovalevsky's arrival coincided with Sofya's most intensive work on her project, endured Kovalevsky's presence for ten days and then persuaded him to remove himself to Uppsala for a while. Sofya wrote to Anne-Charlotte in Italy:

> M. left yesterday evening. . . . if he had stayed here I don't really know whether I would have been able to complete my work. He is so big . . . and takes up so terribly much room, not only on the couch, but also in people's thoughts, that it would have been truly impossible to think of anything else but him in his presence. . . . I am still quite incapable of analyzing my feelings toward him. . . . He's a real Russian from head to foot. And it's also true that he has more intelligence and originality in his little finger than you could extract from both Messrs. X together, even if you put them under a hydraulic press. (VIP, 301–302)

He proposed marriage to her, but there was a condition attached: she would have to give up her work. Even if she had been willing to do that, she could not back out of the prize competition at that point; her involvement was already known to all her colleagues. She lived through the months before the deadline in an almost schizoid state, working to exhaustion and tormenting herself with the suspicion that he was growing cool to her.

Of the fifteen entries anonymously submitted to the Academy of Sciences, one was found to be so significant a contribution toward the general solution of the problem that the prize money was increased from 3000 to 5000 francs; and when the sealed envelope accompanying the work was opened, the author was revealed as Kovalevskaya.[21] At the end of December, 1888, accompanied by a sober, withdrawn Maxim Maximovich, she went to Paris to receive her award. She was the woman of the hour, so lionized that the presence of the man beside her was barely noticed. As soon as the festivities were over Maxim Maximovich left by himself for his Riviera villa at Beaulieu.

It was the actualization of Kovalevskaya's often-repeated lament that life had presented her with everything she had ever desired, but always at the wrong time and under circumstances that converted her happiness into misery. From then to the end of her life, the relationship with Maxim Maximovich was an unbroken alternation of quarrels and reconciliations.[22]

The inability of the pair either to break off their relationship or to formalize it by marriage undoubtedly contributed to Kovalevskaya's growing discontent with her life in Sweden during this period. Realizing that the term of her professorial appointment was due to expire in 1889, she made repeated efforts to find a teaching position in Russia or, failing that, in France. Her election as corresponding member of the Russian Academy of Sciences raised her hopes for a time. But a letter on the subject from K. S. Vesleovsky, permanent secretary of the Academy, made her true position clear:

> Inasmuch as faculty positions in our universities are completely closed to women, no matter how great their ability and learning, there is no place in our country for Madame Kovalevskaya which would be as distinguished and remunerative as the one she now occupies in Stockholm. The position of mathematics teacher in the Higher Courses for Women is considerably lower than a chair at a university; and in other institutions of learning in which women may teach, instruction in mathematics is limited to the elementary level. (VIP, 353–354)

In France, the doors were closed as well, despite Kovalevskaya's high connections and the many efforts put forth on her behalf. She therefore gave up her illusions of living in closer proximity to Maxim Maximovich and accepted a lifetime appointment in Stockholm.

At the very beginning of the relationship with Maxim Maximovich Sofya had told Anne-Charlotte that in his presence she was fired with the urge to write. The next two years were in fact extraordinarily productive. *Memories of Childhood* began, according to Sofya's own account, as a casual vacation pastime at the Beaulieu villa: she happened to recount some episodes from her childhood which Kovalevsky and his friends found interesting, and they urged her to write them down. But it is evident that the impulse went much deeper and represented a major effort to retrieve and understand the past. As she had written to her cousin after Anyuta's death, "There is no one left for whom I can be that bashful, diffident, clinging little Sonya."

The writing went with great speed. The first draft was done in a little over a month. During the same vacation she also tossed off a reminiscence of the Polish uprising of 1863, which (to her later regret) she wrote in French because she knew that the censorship would make its publication in Russia impossible.

Back in Stockholm for the beginning of the academic year, she finished the manuscript within the next few months. A translator was found to put it into Swedish, and it was read aloud, chapter by chapter, to a group of close friends. In deference to the easily shocked sensibilities of Swedish society, it was decided to issue the book as a novel with the title *The Rajevski Sisters.* It was an enormous success. It appeared in Russia in its original form during the following year.

In the same period, Kovalevskaya finished the prologue and part of the first chapter of a novel she had been thinking about since 1886: *Vae Victis* (Woe to the Vanquished). Of the novel itself we know only that it was intended to convey something of its author's own inner life. The prologue, which was published in Swedish and in Russian, is an anti-apotheosis of spring, depicting it not in conventional poetic terms but as a crude sensual force which awakens desire and seems to offer a mass of promises but fulfills none of them.

During that same extraordinarily productive autumn Sofya and Anne-Charlotte embarked on their second and last collaborative venture. This was a play intended for the Swedish stage, based on an incomplete manuscript of Anyuta's, *Before and After Death.*

In 1890 Kovalevskaya wrote *A Nihilist Girl,* her radical *Bildungsroman.* Since it was intended for foreign consumption (there being no possibility of its being published in Russia) the first draft was written in Swedish. She wrote another draft in Russian but died before finishing the final version; the final editing was done by Maxim Kovalevsky on the basis of the two existing drafts, and the book was published posthumously in Geneva.[23]

The body of journalism produced by Kovalevskaya from 1888 to 1890 bears witness to the breadth of her interests. An article on Sweden's peasant universities[24] should be mentioned, as well as two articles of exceptional interest on the controversial new technique of treating hysteria by hypnosis. One of these discusses a psychiatric session led by Dr. Louis at the Paris hospital La Charité, the other gives an account of a clinical lecture by Charcot in the neuropsychiatric clinic he founded at La Salpêtrière.[25]

At the time of her death, Kovalevskaya had several major projects partly finished or still in the planning phase: a novel in which Maxim Kovalevsky was the central figure; a novel on Chernyshevsky; *Vae Victis;* a series of stories and sketches of French life; and a sequel to *Memories of Childhood* which was to deal with the Krukovsky sisters' experiences in the Commune of Paris in 1871.[26]

There is an interesting letter of Kovalevskaya's written in

1890 and quoted here in part:

> . . . you are surprised at my working simultaneously in literature and in mathematics. Many people who have never had occasion to learn what mathematics is confuse it with arithmetic and consider it a dry and arid science. In actual fact it is the science which demands the utmost imagination. One of the foremost mathematicians of our century says very justly that it is impossible to be a mathematician without also being a poet in spirit. It goes without saying that to understand the truth of this statement one must repudiate the old prejudice by which poets are supposed to fabricate what does not exist, and that imagination is the same as "making things up." It seems to me that the poet must see what others do not see, must see more deeply than other people. And the mathematician must do the same. (VIP, 314)

She would doubtless have been the first to characterize her death as a piece of needless stupidity. Over the Christmas holidays she had gone to Kovalevsky's Riviera villa in Beaulieu in the joyful expectancy that always characterized their reunions at the beginning. She wrote a letter from Beaulieu to Sweden which the recipient, Ellen Key, described as "saturated with sunlight, happiness and fragrance." But the inevitable pattern of jealous scenes and demands repeated itself, and by the arrival of the New Year the relationship was back on its old wretched footing.

Before Christmas Sofya and the newly married Anne-Charlotte, who had never met Maxim, had arranged to spend New Year's day together in Genoa. Because of a misaddressed telegram the meeting never took place, and Sofya and Maxim spent the day alone in a Genoa cemetery, wandering among old gravestones and rehashing old differences.

As a result of her heedlessness of practical details, the long journey back to Stockholm turned into a nightmare of errors. She arrived in Denmark in the middle of the night with no Danish money for a porter and had to carry her own baggage in the pouring rain. Although she was already ill, she forced herself to teach her class and then to attend a dinner party. Not until she was in a state of near-collapse did she send for a doctor, who misdiagnosed her illness as "kidney colic." By the time the diagnosis of "suppurative pleurisy," i.e. pneumonia, was made, the disease had progressed with terrible rapdity; in six days she was dead. On the last day of her life the doctor expressed the opinion that the danger had passed; and Fufa, then twelve, was sent off to a children's ball.

The funeral was attended by a mass of luminaries. Fritz Läffler read a poem; Maxim Kovalevsky arrived in time to deliver the eulogy. Wreaths and telegrams poured in from the far corners of the Russian Empire, from students in Tiflis and schoolmistresses in Kharkov. Pyotr Lavrov read a speech at a memorial meeting in Paris in which he spoke less about Kovalevskaya than about his concept of woman's vocation.[27] The Russian minister of the interior, I. N. Durnovo, expressed the opinion that entirely too much attention was being paid to "a woman who was, in the last analysis, a Nihilist."[28]

During the autopsy her brain was uncovered, as was the custom of that era when brain weight and formation were considered the index of intellect; and the Stockholm newspapers ran the information that the "brain of the deceased was developed in the highest degree and was rich in convolutions, as might have been predicted, judging by her high intelligence."[29]

At first glance, *A Russian Childhood* belongs securely in the tradition of family chronicle so well known to Russian literature; but this appearance is deceptive. Except for the "Palibino" chapter, evidently intended for foreign consumption, the book is no cozy genre painting. Memory here, unembellished by lyrical evocation or idealization of the past, is associated less with nostalgia than with identity itself: the attempt to retrieve and understand the process by which a sense of self is formed, from the child's first germinal awareness of its own "I," through evolving images of self as defined in relationships with others within the family setting, up to adolescence's explosion into the world outside.

Within this matrix Palibino and its dramatis personae unfold in the sequence and dimensions of their impingement on the child Sofya: *nyanya,* sister and brother, parents, teachers, a pivotal pair of uncles, the coexisting world of servitors and serfs. These, once sprung out of the narrator's memory, immediately assume that vivid objective reality which is the property of imaginative literature at its best, independently of documentary accuracy. This fact doubtless explains the enormous success of the book when it was first published in Sweden as a work of fiction by transferring the narration from first to third person.

Kovalevskaya's prose style is notable for extreme simplicity and compression, qualities for which her mathematical work was also renowned. The narrative voice is low-keyed, almost neutral, neither self-justifying nor self-lacerating.

What we are given in effect is a double view of reality. On one hand, the very graphic exterior reality of Palibino society—backstairs and kitchen as well as nursery and drawing room—conveyed in the context of the ongoing domestic dramas being played out, and with a sharp ear for individual cadences and levels of speech. On the other, the data filtered through the consciousness of a child and interpreted in accordance with her limited knowledge and large emotional capacity. Very early in her life Sofya came to believe that she was not loved; and that primal feeling colored every subsequent encounter, no matter how diverse, with other human beings. The slightest indication from others of preference for her evoked a wildly exaggerated response bordering on adoration. And at the same time she began to use her extraordinary intelligence in the service of a hopeless endeavor: to extract from others by intellectual achievement the love not hers by virtue of her simple existence.

Indeed, the power of varieties of love to shape human personality is the unifying theme of the book. It is interesting that the child Sofya's earliest memories relate not to her mother but to *nyanya,* that loving, enveloping presence (her metaphor is her own bed in the nursery piled with featherbeds) for whom Sofya was the special favorite of the three children.

If the metaphor for *nyanya* is a pile of featherbeds, then Miss Margaret Smith, the English governess into whose care Sofya next passed, might be visualized as a metal ruler: something rigid and straight that takes measurements. It was her objective to transform the Russian eight-year-old into what she conceived to be a "proper English miss," that is to say, a child from whom the curiosity, exuberance, playfulness—in a word, the childhood—have been expunged. She was deeply antagonistic to any kind of uncensored learning or independent exploration; so she drove Sofya into reading clandestinely. She removed the pleasure even from play by making it compulsory.

In the six years of Miss Smith's residence in the Krukovsky household, perhaps the most destructive feature of her pedagogic practice—certainly the one most hateful to Sofya—was her calculated policy of isolating the child from the rest of the family and particularly from Anyuta, whose vibrancy and physical beauty, independence and advanced social views made her ideologically unacceptable and personally abhorrent to the governess.

And yet, strangely enough, when Miss Smith was finally dismissed from her post Sofya was unable to feel the pleasure which would have been the appropriate response to her release from a person she disliked and feared. On the contrary, a single and entirely uncharacteristic gesture of affection from the glacial governess threw her into a state of emotional upheaval. "I was overwhelmed by piercing anguish, by a sense of irretrievable loss, as if our whole family were disintegrating with her departure." The child's hunger for love was by now so desperate that she was ready to barter her freedom and the integrity of her own feelings to keep it. The episode is a remarkable insight into the ways in which child personality can be distorted when denied satisfaction of primary needs.

It is interesting that the book contains no mention of Kovalevskaya's mathematical gifts, although these broke forcibly upon the family's attention when she was twelve or thirteen years old. What is treated is the genesis of her attraction to the philosophical aspect of mathematics, developed, as she describes it, through a loving relationship with an adored, eccentric uncle under whose influence she began to look on abstract thought in general, and mathematics in particular, as a kind of enchanted kingdom which might be entered if the formulas for admission could only be deciphered. Put into this context, the book supplies another clue to the emotional function of mathematics in her life.

What little of *A Russian Childhood* is generally familiar to readers deals with Dostoevsky's courtship of Anyuta. Even this acquaintance is very likely second-hand (via a literary historian who paraphrases Kovalevskaya's account line for line) or inaccurate (stemming from Anna Grigorievna Dostoevskay's memoirs of her husband, which repeat his fable that he broke the engagement with Anyuta because of ideological differences).

The portrait of Dostoevsky given in the book is brilliant and has undoubted documentary value. Within the narrative framework, however, the literary persona Dostoevsky performs another function: the dramatic role of touchstone against which each of the two sisters, Sofya and Anyuta, tests her conception of what a woman is. This concurrent growing up is beautifully worked out. Maintaining her double viewpoint with great control, the author shows Anyuta's coming of age as a series of chrysalis phases out of which she emerges each time with a new experimental identity (queen of Petersburg society, heroine of a courtly romance, Christian martyr, radical atheist, and finally writer); and also as it registers on the adolescent Sofya, trying to deal with her own complicated feelings about her sister and herself.

The three-month romance of Anyuta and Dostoevsky in Petersburg in the winter of 1865 begins to take on an almost Dostoevskian coloration as it progresses from its first phase (in which Anyuta is filled with awe of her literary idol) to a stage of warm family friendship (in which Dostoevsky freely discusses his epilepsy, his work and some of his thematic preoccupations, including the rape of a ten-year-old girl) to a final stage in which the power relations between the two are reversed. This reversal is precipitated by an embarrassing scene at a Krukovsky soiree during which Dostoevsky's clumsy possessiveness toward Anyuta culminates in his humiliation in the presence of all the guests (the prototype, as has been pointed out, of the celebrated scene in *The Idiot* where Myshkin, a guest in the Epanchin drawing room, launches into a tirade, breaks a vase and has an epileptic fit).[30] After that evening the relationship changes into a sado-masochistic courting game in which Anyuta easily maintains the upper hand and in which radical ideology is used as a screen for deeper quarrels. Dostoevsky would accuse Anyuta of holding boots in higher esteem than Pushkin, and she, knowing the precise response guaranteed to infuriate him, would agree that Pushkin was *irrelevant.*

In these skirmishes Sofya, who was then fifteen, became

a pawn because she was too young to understand courting games. She was already deeply infatuated with Dostoevsky and was living in a state of intense adolescent fantasy and longing. So that when he began to make a practice of comparing the intellectual, spiritual and even physical endowments of the two sisters, much to the younger's advantage (Sofya was the sensitive one, she understood him, and so forth), she took his words in dead earnest. She began to believe he really did prefer her to Anyuta.

When she overheard a conversation in which Dostoevsky confessed to Anyuta that he loved her, not in friendship but in passion and with his whole being, and begged her to become his wife, Sofya experienced pain of a truly adult magnitude. It was the old sensation of cosmic loneliness (the aftermath of all her emotional overinvestments) mixed with bitterness and embarrassment at having been made a fool of. It was not until days later that she learned that Anyuta had in fact declined Dostoevsky's proposal. Anyuta's explanation of why she did this bears repeating in the present context because of its simple and penetrating formulation of the identity problem:

> He needs an entirely different kind of wife from me. His wife will have to dedicate herself to him utterly, utterly, to give her whole life to him, to think about nothing but him. And I can't do that, I want to live myself! . . . He always seems to be taking possession of me and sucking me up into himself. When I'm with him, I can never be myself.

The Dostoevsky episode is both climax and denouement. The narrative ends on a lyrical note, one of the few in the book. Madame Krukovskaya and her daughters are traveling the melting roads from Petersburg toward home at the beginning of spring. The experience with Dostoevsky has been catalytic, but differently for each of the sisters: for Anyuta, crystallization of her sense of herself, for Sofya the disclosure of emotions she does not know how to handle:

an unsettling pre-vision of adulthood. But all of this is swallowed up for a little while in the April night, in the passage of the carriage through the pine woods toward home, while the sisters cling to one another in the unreasoning happiness of youth and the blind conviction that the life stretching out ahead of them will be boundlessly long and good.

At the time when Anyuta was explaining to her younger sister that she could not think of marrying Dostoevsky because her identity disappeared in his presence, Sofya had a distinct reaction of her own: "God, what happiness to be with him constantly, and submit yourself to him utterly! How could my sister push such happiness away?"

It was an extraordinary revelation, a key to one of the major discords of her life: on the one hand, a driving need to keep her individuality intact and develop her creative capacities to their limits; on the other, an oceanic hunger for love which she understood as fusion with the personality of the loved one. The confusion was further compounded by a society which regarded it as woman's obligation to give herself to her husband indivisibly, retaining to herself no private sphere of thought or act.

The temptation is irresistible to juggle the data of time and place a bit for speculative purposes. What might have been different in Kovalevskaya's life if, say, she had been born in the United States about a century later? There would then have been no need for her to explain her motivation to teach in an institution of higher learning in order to open up university admission to women once and for all. "As things stand now, that admission is either an exception or a special favor which can always be revoked." (VIP, 258).

Her acceptance at every stage of the academic process would be assured. She would not have to be a near-genius to enter graduate school. No inordinate obstacles would be put in her way, either in earning her doctorate or in university teaching (although it is just possible that her academic advancement might come to a halt at the level of assistant professor). Her life would not have been choked off at forty-one, because antibiotics would have made her death unnecessary. She would not have been driven into a fictitious marriage, but would have been free to choose from a whole spectrum of options: traditional marriage, "open" marriage, group marriage or no marriage at all. But in living through the overriding question of her life—what is love, and how should it mediate with freedom, identity, commitment—she might find contemporary society not very much further along the way to a solution than hers was, when she took her first step toward independence in 1868.

NOTES

1. S. V. Kovalevskaya, *Vospominaniya i pis'ma,* ed. S. Ya. Shtraikh (M, 1951), p. 485. Hereafter referred to as VIP, followed by page numbers in the text.

2. Since Russian law made it almost impossible to obtain a divorce.

3. His later work laid the foundation for the study of evolutionary paleontology in Russia. His brother, Alexander Onufrievich Kovalevsky (1840–1901) was a distinguished biologist and a pioneer in the study of comparative embryology.

4. Ivan Mikhailovich Sechenov, 1829–1905. Eminent physiologist, professor at the University of Petersburg.

5. N. G. Chernyshevsky, *Chto delat'?* (1864). Maria Bokova-Sechenova is generally regarded as the model for the character Vera Pavlovna, Bokov for Lopukhov, Sechenov for Sasha Kirsanov. The assumption is now disputed by some Soviet scholars.

6. Anna Mikhailovna Evreinova, 1844–1919. Wrote and lectured on women's rights and later edited *Severny vestnik* (The Northern Messenger).

7. I. S. Knizhnik-Vetrov, *Russkiye deyatel'nitsy pervogo internatsionala i parizhskoy kommuny* (M, 1964), 179.

8. Ibid., 192–193.

9. The rest of their life together was a tragic history of poverty, illness, rootlessness, exile. Anyuta later published two novellas but never fulfilled the promise of her youth.

10. 770 people were arrested, of whom 258 were kept in prison for years before trial. For excellent description of the movement see F. Venturi, *Roots of Revolution* (New York, 1966), Chapter 18.

11. L. A. Vorontsova, *Sofya Kovalevskaya* (M, 1957), 147.

12. 1876 and 1877.

13. Vorontsova, 156.

14. See her poem, "Zhaloba muzha," (A Husband's Complaint) in S. V. Kovalevskaya, *Vospominaniya i povesti* (M, 1974), 319.

15. Litvinova, 58. Litvinova adds a perceptive comment: "Such an attitude toward herself explains a great deal about her career. . . . Like all the prejudices we must struggle against, the prejudice against the ability of women to do intellectual work exists not only in those around us, but also in ourselves. . . . It would never occur to the most mediocre man that he was not sufficiently prepared to carry out the duties of a lecturer."

16. "Vospominaniya o Dzhorzhe Elliot," *Russkaya mysl',* No. 6 (1886).

17. In Swedish, *Kampen för lyckan.*

18. It was staged several times in Moscow, apparently with much success.

19. M. M. Kovalevsky, 1851–1916. He was a distant relative of Sofya's husband. After his dismissal from Moscow he lectured at Oxford, Paris and Chicago. Wrote prolifically, including a memoir of Marx. After 1905 returned to Russia, resumed teaching and served in the State Duma.
20. Vorontsova, 267.
21. Kovalevskaya's own account of the event in *An Autobiographical Sketch,* 225–227.
22. Although there is evidence that they were planning to be married in the summer of 1891.
23. It appeared in Swedish, French, German, Polish, Czech and English, but was repeatedly prohibited by the Russian censor up to 1906, when an edition appeared together with a note by Kovalevskaya's daughter that the honorarium had been contributed by the author's estate to a fund in aid of political prisoners. The next Russian edition was published in 1922.
24. "Tri dnya v krest'yanskom universitete v Shvetsii," *Severny vestnik,* No. 11 (1890).
25. "V bol'nitse La Charité" and "V vol'nitse La Salpêtrière," *Russkiye vedomosti,* Oct. 28, No. 1, 1888, under the pseudonym of Sophie Niron.
26. The chief authority for this is Ellen Key, with whom Kovalevskaya discussed her literary plans during the last days of her illness.
27. Pyotr Lavrov, *Russkaya razvitaya zhenshchina* (Geneva, 1891).
28. Vorontsova, 333.
29. Reprinted in *Novoye vremya,* February 13, 1891.
30. Just as it has been pointed out that Anyuta herself was the prototype for Aglaya Epanchina (for example by E. H. Carr, *Dostoevsky* (London, 1962), 101, 167.

CHAPTER ONE

Earliest Memories

I would like to know whether there is anyone who can pinpoint that precise instant of existence when a clear awareness of his or her own "I" emerged for the first time: the earliest glimmer of conscious life. I cannot do it at all. When I start sorting through my first memories and classifying them, the same thing happens to me every time: these memories always seem to slide apart before my eyes. Here it is, it seems—I've found it, the first impression that left a distinct memory trace. But no sooner do I focus my thoughts on it for a while than other impressions from an even earlier time immediately appear and take form.

And the most troublesome thing about it is that I myself am utterly unable to determine which of these impressions I actually remember from experience and which of them I only heard about later in my childhood; so that I imagine I remember them, whereas all that I remember in actual fact is other people's accounts of them. And what is worse—I can never manage to call up a single one of these primal memories in all its purity without unwittingly mixing something alien with it during the actual process of remembering.

Be that as it may, here is one of the first pictures which comes into being when I begin to recall the very first years of my life.

The pealing of church bells. The smell of incense from the priest's censer. A crowd of people is coming out of church. My nanny is leading me by the hand down from the church porch, solicitously shielding me from being jostled. "Don't hurt the little one!" she incessantly entreats the people crowding around us.

At the church exit an acquaintance of Nanny's wearing a long cassock (a deacon or sexton, probably) comes up to us and offers her the Communion wafer. "Eat it in good health, lady," he says to her.

And then he turns to me. "And you, my clever little one. Tell me, what's your name?"

I don't answer, merely stare at him, all eyes.

"Aren't you ashamed not to know your own name, little miss?" the sexton teases.

"Tell him, honey," Nanny coaches me. "'My name is Sonechka,' tell him, 'and my Papa is General Kryukovskoy.'"[1]

I try to repeat it after her, but it evidently comes out all mixed up, since Nanny and the man both laugh.

Nanny's acquaintance sees us home. I hop about all the way home and go over Nanny's words, twisting them up in my own fashion. Evidently this idea of my own name is still something new to me, and I try to stamp it into my memory.

As we approach our house the sexton points to the gates. "Look there, my little miss, there's a *kryuk,* a hook, hanging on the gate. When you forget your Papenka's name all you have to do is think, 'There's a *kryuk* on the Kryukovskoys' gate.' And then you'll remember it right away."

And so it was. Embarrassed as I am to admit it, that bad

pun of the sexton's remained with me and initiated an era of my existence. From it my chronology begins, the earliest appearance inside me of a distinct self-awareness: who I am, what is my place in the world.

As I think back over it now, I believe that I must have been two or three years old at the time, and that the episode took place in Moscow, where I was born. My father served in the artillery and we often had to move from city to city, following after him as his work demanded.

After this first, clearly preserved scene there is another long gap, against whose misty gray background different little scenes of travel stand out like bright, scattered specks: gathering pebbles along the highway, stopping overnight at post-stations, my sister's doll thrown out of the carriage window by me—a series of random but fairly vivid pictures.

Memories with some degree of coherence did not begin until I was five, when we were living in Kaluga.[2] There were three of us children by then. My sister Anyuta[3] was about six years older than I, and my brother Fedya[4] about three years younger.

I see our nursery very clearly. A big but low-ceilinged room. If Nanny climbs up on a chair she can easily touch the ceiling with her fingers. All three of us sleep in the nursery. There has been some talk of moving Anyuta to her French governess's room to sleep, but she doesn't want to do it, preferring to be with us.

Our little beds, fenced all round with slats, stand side by side, so that we can climb into each other's beds in the morning without touching our feet to the floor. Nanny's big bed stands some distance apart, with a whole mountain of featherbeds and quilts piled on top of it.

This bed is Nanny's pride. Sometimes during the day, when she is in a good mood, she lets us wallow in her bed. We clamber up on it with the help of a chair, but the moment we reach the very peak the mountain collapses under us, and we're buried under a soft sea of down. This gives us enormous amusement.

No sooner do I think of our nursery than immediately, by an inescapable chain of associations, I begin to perceive a certain smell. It is a mixture of incense, low-grade oil, May balsam, and the smoke from a tallow candle. It has been a long time since I have had occasion to smell that special smell. Indeed, I would imagine that one would encounter it seldom, not only in Europe but even in Petersburg or Moscow. But a few years ago, when I was visiting friends in the country, I went into the nursery and was immediately met by that familiar smell, which evoked a whole train of long forgotten memories and sensations.

The governess, a Frenchwoman, could never enter our nursery without putting her handkerchief to her nose squeamishly.

"Do open the transom, Nanny!" she would implore in her broken Russian.

Nanny took this comment as a personal insult.

"What's she got into her head now, the heathen? I can just see myself opening up the windows so the master's children can catch cold!" she would grumble after her as she left.

Nanny's skirmishes with the governess were repeated regularly every morning.

The sun has been peeping into the nursery for a long time. One after another, we children open our eyes, but we don't hurry to get out of bed and get dressed. Between the time of awakening and the time of starting to dress there is still a long interval ahead of us: making a racket, throwing pillows at each other, grabbing each other by our bare feet, prattling every imaginable kind of nonsense.

An appetizing aroma of coffee spreads through the room. Nanny, who is only half dressed herself (having merely exchanged her nightcap for the silk kerchief which invariably covers her hair during the day), comes in carrying a tray with a big copper coffeepot on it. She begins to feed us—still in our beds, still unwashed and uncombed—coffee with cream and rich fat buns.

It sometimes happens that after breakfast, tired out by all the preliminary to-do, we fall asleep again. But now the nursery door opens with a bang, and the furious governess appears on the threshold.

"What! You are still in bed, Annette! It is eleven o'clock! You are late for your lesson again!" she angrily exclaims in French. And adds in broken Russian for Nanny's benefit, "To sleep so long is not allowed! It will be to complain to the General!"

"Go then, you snake—go and complain!" Nanny mutters after her. After the governess has left, Nanny cannot calm down for a long time and keeps grumbling, "So now the master's children aren't even allowed to have a little sleep! Late for your lesson! What's the great harm in that? All right, you can just wait a while—it's nothing to be so stuck-up about!"

All the same, grumbling or no, Nanny now considers it in order to settle down seriously to the business of getting us washed and dressed. I must confess that if the preparations were lengthy, the toilette itself is accomplished with great speed. Nanny wipes our faces and hands with a wet towel, passes a comb once or twice through our tousled manes, dresses us in dresses which are often missing a few buttons—and there it is, we're ready!

My sister sets off to her governess for her lesson, while my brother and I stay in the nursery. Not constrained by our presence, Nanny sweeps the floor with a brush, raising a cloud of dust. She smooths the blankets over our cribs and shakes out her own featherbeds. It is now considered that the nursery is tidied up for the day.

My brother and I sit on the oilcloth couch (which is torn in several places, so that the horsehair sticks out of it in big clumps), and we play with our toys. We are seldom taken out for walks except when the weather is especially fine, and also on the major holidays when Nanny takes us to church.

When her lesson is over, my sister immediately comes running to us again. She finds it boring with her governess. Things are livelier with us, particularly because visitors often come to see our nanny—other nannies or housemaids to whom she serves coffee and from whom one can hear lots of interesting things.

Sometimes Mama looks into the nursery. When I recall my mother during that first phase of my childhood, I always picture her as a quite young and very beautiful woman. I see her always gay and elegantly dressed. Most often I remember her in a low-necked ball gown, with bare arms, and wearing a mass of bracelets and rings. She is getting ready to go out somewhere to a party and has dropped in to say goodnight to us.

The moment she appears at the nursery doors Anyuta rushes straight to her and starts kissing her hands and neck and examining and fingering all her golden trinkets.

"I'm going to be just as beautiful as Mama when I grow up!" she says, fastening Mama's adornments on herself and standing on tiptoe so as to see herself in the little mirror hanging on the wall. This gives Mama a great deal of amusement.

Sometimes I too feel the desire to snuggle up to my mother, to climb on her lap. But these attempts, somehow, always end with some kind of clumsiness on my part. Either I hurt my mother or I tear her dress, and then I feel ashamed and run away and hide in the corner. For this reason a kind of sullen shyness towards my mother has begun to develop in me, and this shyness is increased by the fact that I have often heard Nanny say that Anyuta and Fedya are Mama's favorites, while I am—not loved.

I do not know whether or not this was the truth. But Nanny repeated it often, oblivious of my presence. Perhaps it merely seemed that way to her precisely because she herself loved me much more than she loved the other children. Although she was the one who raised all three of us, she always considered me her special charge for some reason, and therefore she used to take offense on my behalf for every insult which, in her opinion, was proffered to me.

Anyuta, as the eldest by far, naturally enjoyed greater privileges than we did. She grew up independent as a Cossack, refusing to acknowledge any authority over herself. She was given free access to the drawing room. From early childhood she earned the reputation of being a charming child and grew accustomed to entertaining the guests with her witty and sometimes impertinent remarks and pranks.

My brother and I, on the other hand, appeared in the front rooms only on special occasions. Normally we lunched and dined in the nursery. Sometimes when there were dinner guests Mama's chambermaid Nastasya would come running into the nursery during the dessert course.

"Nanyushka, dress Fedenka right away in his little blue silk shirt and bring him into the dining room! The mistress wants to show him to the company!" she would say.

"And what am I supposed to dress Sonechka in?" Nanny would ask in an angry voice, since she already foresaw what the answer would be.

"You don't have to dress Sonechka in anything, let her sit in the nursery! She's our homebody," the chambermaid would reply with a hearty laugh, knowing that this answer was sure to infuriate Nanny.

And she was right. This wish to exhibit Fedya alone to the guests was regarded by Nanny as a deadly insult to me, and for a long time afterward she would feel resentful and would mutter under her breath, gazing at me with compassionate eyes, patting my head and repeating over and over, "My poor little thing, my little darling."

It is evening. Nanny has already put my brother and me to bed, but she herself has not yet taken off her everlasting silk kerchief, the removal of which signifies her transition from wakefulness to rest. She is sitting on the couch at a round table and drinking tea in the company of Nastasya.

The nursery is half dark. Out of the dimness emerges, like a yellow stain, only the sooty flame of the tallow candle, which Nanny keeps forgetting to trim. In the opposite corner of the room the wavering bluish flame of the icon lamp traces fanciful patterns on the ceiling and vividly illumines the Savior's hand offering a blessing, standing out brightly from its silvered mounting.

I hear the measured breathing of my sleeping brother quite close to me, and from the corner beyond, the stove carries the heavy nasal whistling of the little girl who has been attached to the nursery as a servant—snubnosed Feklusha, Nanny's scapegoat. Feklusha sleeps right here with us in the nursery—on the floor, on a piece of thick gray felt matting which she spreads out in the evening and keeps hidden in a closet during the day.

Nanny and Nastasya are talking in an undertone. Imagining that we are sound asleep, they review all the household events without any constraint. But I am not asleep; on the contrary, I am listening attentively to every word they say. There is much, of course, that I do not understand and much that is simply boring. There are times when I fall asleep right in the middle of some story without hearing it through to the end. But those fragments of their talk which reach my consciousness pattern themselves there in fantastic images, and leave their ineradicable trace for the remainder of my life.

"So then, how can I help loving her better than the other children, my poor little dove?" I hear Nanny saying, and I realize that they are talking about me.

"Wasn't I the one who had to bring her up practically all by myself? The rest of them didn't want to be bothered with her. When we had our Anyutochka, now, her Papenka and her Mamenka and Grandpapa and all her aunties simply adored her, because she was the first. I could never even get to rock her—every minute it was this one or that one, coming to take her away from me! But things were altogether different with Sonechka."

At this point in the story, so often repeated, Nanny always lowers her voice mysteriously. This, it goes without saying, makes me listen even harder.

"She was born at the wrong time, my little dove, and that's the truth of it!" Nanny says in a half whisper. "The master, practically on the day before she was born, he goes and gambles his money away at the English Club . . . drops it all . . . they had to go and pawn the mistress's diamonds! Well then, there were other things on their minds besides being glad that God sent them a daughter! And on top of that, the master and the mistress, both of them wanted a son so much. The mistress kept telling me, 'You'll see, Nanny, it will be a boy!' They went and got everything ready for a boy, right and proper—a little crucifix and a little cap with a blue ribbon on it—and then it didn't turn out that way, what can you do! It had to go and be another girl. The mistress was so upset, she didn't even want to look at her. It was only later, when Fedya came, that he consoled them."

This story was repeated by Nanny with great frequency and I listened to it with the very same curiosity each time, so that it became firmly fixed in my mind. Because of stories like these I early formed the conviction that I was not loved, and this belief affected my entire personality. I grew ever shyer and more withdrawn.

I would be brought into the drawing room, I remember. I would stand there scowling and clinging to Nanny's dress with both hands. It was impossible to get a word out of me. No matter how Nanny tried, I remained stubbornly silent and merely glared at them all, scared and resentful like some little badgered animal; until my mother would finally say with annoyance, "All right, Nanny, take your savage back to the nursery! She's only an embarrassment in front of our guests. She has probably swallowed her tongue!"

I shield away from strange children as well and, indeed, it was only on rare occasions that I saw any. Although I remember that when Nanny and I were out walking and used to see the local boys and girls playing some noisy street game, I often envied them and wanted to join in their play. But Nanny never permitted it.

"What are you thinking of? You, a well-born young lady, you want to play with ordinary children?" she would say in such a convinced, reproachful voice that—I still remember it—I immediately felt ashamed of my wish.

Soon I lost even the desire and the ability to play with other children. I recall that occasionally, when some little girl of my own age was brought to visit me, I would never know what to say to her—I would just stand there and think, will she go away soon?

And so I was all the happier when I was left alone with Nanny. In the evenings, after Fedya had been put to bed and Anyuta had run to the grownups in the drawing room, I used to sit next to Nanny on the couch, nestled up very close to her, and she would tell me fairy tales. The force of their impact on my imagination I assess by the fact that I can remember only fragments of them when I am awake.

But when I am asleep I still to this day dream about them occasionally: sometimes "the black death," sometimes "the werewolf," sometimes "the twelve-headed dragon." And these dreams always call up in me the same unaccountable, breath-stopping horror I felt when I was five years old, listening to Nanny's fairy tales.

At the same period of my life, something strange began happening to me. There were times when a feeling of inexplicable depression came over me—an *angoisse.* I vividly recall the sensation. It usually happened if I was alone in a room when twilight was coming down. I would be playing busily with my toys, thinking of nothing in particular. Suddenly I would look around and spy a sharp black strip of shadow in back of me, crawling out from under the bed or out of the corner. The sensation that came over me then was as though some alien thing had crept into the room, unseen. And because of the presence of this new, unknown thing, my heart would constrict so painfully that I would rush headlong to look for Nanny, whose presence was normally able to calm me down. But there were times when this tormenting sensation did not pass for a long time, perhaps for several hours.

I believe that many nervous children experience something similar. In such cases people usually say that the child is afraid of the dark, but this expression is quite wrong. In the first place, the feeling experienced is very complex and is much more like depression than fear. Secondly, it is evoked not by actual darkness nor by ideas associated with darkness, but precisely by the sensation of *oncoming* darkness.

I also remember a very similar feeling coming over me as a child under quite different circumstances—if, for instance, I was out walking and suddenly saw before me a big, unfinished house with naked brick walls and empty spaces instead of windows. I sometimes experienced the same sensation in summertime also, when I would lie down on my back on the ground and look up at an empty sky.

I began to show other symptoms of extreme nervousness as well—a revulsion bordering on horror, for example, toward any kind of physical deformity. If anyone said anything in my presence about a two-headed chicken or a calf born with three paws, I would shake all over. And then, that night, I would invariably dream about the monster and wake Nanny up with a piercing shriek. To this day I can remember a three-legged man who used to shadow me in my dreams all through my childhood.

Even the sight of a broken doll terrorized me. If I happened to drop my doll, Nanny had to pick it up and report to me whether its head was intact. If not, she had to take it away without showing it to me. I still remember an episode when Anyuta once caught me alone, away from Nanny. Wanting to tease me, she began pushing in my face a wax doll with a broken black eye dangling out of its head; and this act brought me to the verge of convulsions.

Altogether, I was well on the way to becoming a morbid, neurotic child. Soon afterward, however, my whole environment changed, and everything that had gone before came to an end.

CHAPTER TWO

The Thief

When I was about six years old my father retired from Army service and settled in his family estate of Palibino, in the province of Vitebsk. At that time persistent rumors of an imminent "emancipation of the serfs" were already making the rounds, and these rumors impelled my father to occupy himself more seriously with farming, which up to that time had been given over to a steward.

Soon after our arrival in the country an episode occurred in our household which remained vividly in my memory. Moreover, its effect on everyone else in the house was so strong that it was often recalled afterward. And so my own impressions became intermingled with the subsequent stories about it, and I was no longer able to distinguish one from the other. Therefore, I shall describe this episode as I understand it now.

Various articles suddenly began vanishing from our nursery: now one thing, now another. Whenever Nanny forgot about some article over a period of time and then needed it later, it was nowhere to be found, although she was ready to swear that she herself, with her own two hands, had laid it away in the cupboard or the bureau.

These disappearances were treated rather calmly at first, but when they began to occur more and more often and to include articles of ever increasing value, when a silver spoon, a gold thimble and a mother-of-pearl penknife suddenly vanished in succession, an alarm was raised. It was clear that we had a thief in our house. Nanny, who considered herself responsible for keeping the children's belongings safe, was more upset than anyone, and she resolved to unmask the thief at all costs.

It was natural that suspicion should fall first of all on poor Feklusha, the girl who had been appointed to serve in the nursery. True, Feklusha had been with us for about three years, and Nanny had never noticed anything of the sort in all that time. In her opinion, however, this fact didn't prove a thing.

"Before this, the girl was little and didn't understand the value of things," Nanny reasoned. "But now she's older and smarter. And on top of that, her family lives in the village. So she must be snitching the master's property for them."

Reasoning in this fashion, Nanny reached such a deep inner conviction of Feklusha's guilt that she began behaving toward her with ever greater harshness and severity. And the hapless, intimidated Feklusha, feeling instinctively that she was under suspicion, began to acquire an ever more guilty air.

But no matter how stealthily Nanny watched over Feklusha, she was not able to put her finger on anything specific for a long time. And meanwhile the missing articles did not turn up, and new items kept disappearing. One day Anyuta's money-box, which always stood in Nanny's cupboard and contained about forty rubles (if not more), was gone. The news of this last disappearance reached even my father. He summoned Nanny and gave strict orders that the thief must be found without fail. At this point we all realized that the matter was no joke.

Nanny was desperate. But then one night she woke up and heard something: a peculiar munching sound was coming from Feklusha's corner. Already inclined to suspicion, she stealthily, noiselessly stretched out her hand to a box of matches and lit the candle all of a sudden. And what did she see?

There was Feklusha squatting on her heels and holding a huge jam jar between her knees, stuffing jam into both her cheeks and even wiping up the jar with a crust of bread. I should add that our housekeeper had complained a few days before that jam was disappearing from her pantry cupboard.

To jump out of bed and grab the criminal by her pigtail was, it goes without saying, the work of a single second for Nanny.

"Aha! Caught you, you no-good! Speak up—where did you get that jam?" she shouted in a voice like thunder, mercilessly pulling the girl about by the hair.

"Nanny dear! I didn't do anything wrong, and that's the truth!" Feklusha implored. "It was the seamstress, Marya Vasilievna, it was her gave me the jam last night. But she ordered me not to show it to you."

This explanation appeared to Nanny implausible in the highest degree.

"Well, my dear, anybody can see you don't even know how to tell a lie," she said with contempt. "A likely story . . . when did Marya Vasilievna take it into her head to start treating you to jam?'

"Nanny dear, I'm not lying! It's the God's honest truth. You can ask her yourself. I was heating up her irons for her yesterday, and that's why she treated me to the jam. But she ordered me, "Don't show it to Nanny, or else she'll scold me for pampering you."

"All right then, we'll get to the bottom of this thing

tomorrow morning," Nanny decided. And in anticipation of morning she locked Feklusha up in a dark closet, from which her sobbing could be heard for a long while afterward.

The next morning, the investigation began.

Marya Vasilievna was a seamstress who had been living in our house for many years. She was not a serf but a freewoman and enjoyed greater respect than the rest of the servants. She had her own room, in which she dined on food from the master's table. She held herself very proudly in general and kept apart from all the other servants. She was highly regarded in our house because she was such a complete mistress of her craft. People said of her that she had "golden hands." She was, I imagine, getting on toward forty by then. Her face was thin and sickly-looking, with huge dark eyes. She was homely, but I recall that the grownups always said of her the she looked *distingué,* that "you'd never take her for an ordinary seamstress."

She dressed immaculately and kept her room in perfect order, even with certain pretensions to elegance. There were always pots of geraniums on her windowsill, her walls were hung with cheap pictures and, on the shelf in the corner, various porcelain articles were set out which I highly admired as a child—a swan with a gilt beak, a lady's slipper painted all over with pink flowers.

We children found Marya Vasilievna especially interesting because there was a story connected with her. In her youth she had been a beautiful, strapping young woman, a serf in the household of a certain landowner's widow who had a grown son, an Army officer. This son came home on leave and presented Marya Vasilievna with a few silver coins. By ill luck the mistress entered the serf-girls' room at that very moment, and she saw the money in Marya Vasilievna's hands.

"Where did you get it?" she asked, and Marya Vasilievna took such a fright that instead of answering, she swallowed the coins.

She became ill at once. Her face turned black, and she fell choking on the floor. They barely managed to save her life. She was ill for a very long time, and her beauty and freshness vanished forever. Shortly after this episode the old mistress died, and the young master gave Marya Vasilievna her freedom.

We children were entranced by this story of the swallowed coins, and we often hung around Marya Vasilievna begging her to tell us how it had all happened. She used to visit the nursery rather often, even though she and Nanny were not on the best of terms. And we too loved to run to her room, especially at twilight, when she willy-nilly had to put her sewing aside. She would sit down by the window then and, leaning her head on her hand, would begin singing various sentimental, old-fashioned romances in a plaintive voice: "Among the Even Plains" or "Black Flower, Sad Flower."

Her singing was terribly dismal but I loved listening to it, even though it always made me feel sad afterwards. Sometimes it would be interrupted by terrible attacks of coughing, which had been tormenting her for many years and which threatened to tear her dry, flat chest apart.

When, on the morning after the incident with Feklusha, Nanny asked Marya Vasilievna, "Is it true that you gave the girl some jam?" the seamstress, as might have been expected, responded with an expression of astonishment.

"Whatever have you got into your head, Nanyushka?"

she answered in an offended tone. "Would I pamper the brat like that? Why, I don't even have any jam for myself!"

So now it was all clear. And yet Feklusha's insolence was so great that she went on insisting she was innocent in spite of the seamstress's categorical assertion.

"Marya Vasilievna! As God is watching—did you forget? You called me last night yourself, yes, you did, you praised me for heating up the irons, and you gave me the jam," she kept on repeating in a desperate voice breaking with sobs, and shaking all over as if in a fever.

"You must be sick and raving, Feklusha," Marya Vasilievna answered calmly, her pale, bloodless face betraying no trace of emotion.

And now neither Nanny nor anyone else in the household had any further doubt of Feklusha's guilt. The culprit was taken away and locked into a closet far from all the other rooms.

"Sit there without food or water, you nasty thing, until you confess!" Nanny said, turning the key in the heavy lock.

This event, it goes without saying, raised a commotion all through the house. Every one of the servants thought up some pretext to come running to Nanny to discuss the interesting new development. There was a regular club meeting going on in our nursery all day.

Feklusha had no father. Her mother lived in the village and came to our house to help our laundress with the washing. Naturally, she soon found out what had happened and came at a run to the nursery with noisy and profuse complaints and protestations that her daughter was innocent. But Nanny was quick to quiet her down.

"Don't make such a big noise, lady! Just wait a little bit, and we'll get to the bottom of things, we'll find out where that daughter of yours stashed the stolen goods!" she said so harshly and with such a meaningful look that the poor laundress lost her courage and took herself off.

Popular opinion was decidedly against Feklusha. If she snitched the jam that means she snitched the rest of the stuff too," everyone said. The general indignation against the girl ran particularly high because these mysterious and repeated disappearances had been hanging like a heavy burden over all the servants for many weeks. Each one feared in his heart that he might be suspected, God forbid. Therefore the unmasking of the thief was a relief to everyone.

But just the same, Feklusha would not confess.

Nanny went to visit her prisoner several times in the course of the day, but she kept stubbornly repeating her refrain, "I didn't steal anything. God will punish Marya Vasilievna for harming a fatherless child."

Toward evening my mother came into the nursery.

"Aren't you being a trifle too harsh with the miserable girl, Nanny?" she said with some concern. "How can you leave a child without food all day?"

But Nanny would not hear of clemency. "What are you thinking of, my lady? To take pity on such a one as that! Didn't she almost manage to bring honest people under suspicion, the low, nasty thing!" she asserted with such conviction that my mother was unable to go on insisting and left without lightening the young criminal's lot by one iota.

The next day came. And Feklusha still refused to confess. Her judges were already beginning to feel a certain uneasiness when suddenly Nanny went to see our mother at dinnertime, with an expression of triumph on her face.

"Our little bird has sung!" she said happily.

"In that case," Mama very naturally asked, "where are the stolen things?"

"She still won't tell us where she hid them, the nasty thing!" Nanny replied. "She prattles all kinds of rubbish.

She says, 'I forgot.' But just let her sit under lock and key for another hour or two—and maybe it'll all come back to her!"

And indeed Feklusha made a full confession toward evening, describing in great detail how she had stolen all these articles with the object of selling them later. Since no convenient occasion had presented itself, however, she had kept them hidden for a long time under the thick matting in the corner of her little closet. But then, when she saw that the disappearances had been noticed and that the thief was being hunted in earnest, she got scared. First she thought she would simply put the things back where they belonged, but then she was afraid to try that. So she wrapped them all up in a bundle inside her apron and threw them into a deep pond on the other side of our estate.

Everybody wanted so desperately to find some solution to this painful affair that Feklusha's tale was not subjected to very close scrutiny. After some lamentation over the needless loss of the articles, all satisfied themselves with her explanation.

The culprit was released from detention and a short, just sentence was pronounced over her. It was decided to give her a good hiding and then send her back to the village to her mother. Despite her tears and her mother's protests, this sentence was carried out immediately. Afterwards another girl was sent to serve the nursery in Feklusha's place.

Several weeks passed. Little by little order was restored in the household, and everyone began to forget what had happened.

But then one evening, when everything was quiet in the house and Nanny, having put us to bed, was getting ready to retire for the night herself, the door to the nursery opened softly. The laundress Alexandra, Feklusha's mother, was standing there. She alone had stubbornly resisted admitting the obvious and continued to maintain without surcease that her daughter had been "harmed for nothing." There had already been several strong altercations with Nanny on this point, until Nanny finally gave up and forbade her to come into the nursery any more, deciding that it was useless to try to reason with a stupid peasant woman.

But this time Alexandra had such a strange and meaningful expression on her face that Nanny took one look at her and immediately realized that she was not there to repeat her usual empty complaints, but that some truly new and important event had occurred.

"Now you just look here, Nanyushka—look what a thing I am going to show you," Alexandra said mysteriously. And, looking cautiously around the room to make sure that no outsider was there, she drew out from under her apron and handed over to Nanny a mother-of-pearl penknife—our beloved knife, that very knife supposedly among the stolen loot Feklusha had thrown into the pond.

When she saw the knife, Nanny spread her hands helplessly. "Wherever did you find it?" she asked.

"That's just the point—*where* I found it," Alexandra slowly drawled out her answer. She said nothing for a few seconds, evidently taking pleasure in Nanny's discomfiture. Finally she said ponderously, "That gardener of ours, Filipp Matveyevich, gave me his old pants to darn, and I found the knife inside the pocket."

This Filipp Matveyevich was a German who held one of the leading positions in the servants' aristocracy. He received a rather large salary, was a bachelor, and although to the unprejudiced eye might have seemed no more than a fat German, no longer young and rather repulsive with his typical reddish squared-off sidewhiskers, still our female servants regarded him as a handsome fellow.

Hearing Alexandra's strange testimony, Nanny couldn't take it in for the first minute or two.

"But how could Filipp Matveyevich get hold of the children's penknife?" she asked in confusion. "After all, he practically never goes into the nursery! And anyway, how could it be possible that a man like Filipp Matveyevich would take to stealing things from the children?"

Alexandra gazed at Nanny in silence with a long, mocking stare. Then she bent down right to her ear, and whispered several sentences in which the name of Marya Vasilievna was repeated more than once.

Little by little a ray of light began to penetrate into Nanny's mind.

"Tut, tut, tut . . . so that's how it is!" she said, waving her hands helplessly. "Akh, you humble one, you! Oh, you no-good woman, you!" she exclaimed, filled with indignation. "Just you wait, we'll make you come clean!"

It turned out (as I was later told) that Alexandra had been nurturing suspicions of Marya Vasilievna for a long time and had observed that the seamstress was carrying on a secret love affair with the gardener.

"Well, then," she told Nanny, "judge for yourself. Would a fine lad like Filipp Matveyevich love an old woman like that just for nothing? She was probably buying him with presents."

And indeed she soon became convinced that Marya Vasilievna was giving the gardener both gifts and money. Where then was she getting these things? And so she set up a regular system of espionage over the unsuspecting Marya Vasilievna. The penknife was only the final link in a long chain of evidence.

The story was turning out to be more fascinating and diverting than would have been possible to predict. Within

Nanny had suddenly awakened that passionate detective instinct which so often slumbers in old women's hearts and incites them to rush fervently into investigating all sorts of complicated affairs which do not concern them in the least. And in this particular instance, Nanny's zeal was spurred even more because she felt that she had deeply wronged Feklusha, and she burned with the desire to atone post-haste. Right then and there she and Alexandra formed a defensive and offensive union against Marya Vasilievna.

Since both women were filled with moral certainty of the seamstress's guilt, they resolved upon an extreme measure: to get hold of her keys and (seizing an opportunity when she would be away) to open up her trunk.

The thought is sister to the act. Alas! Their assumptions, as it turned out, were entirely correct. The contents of the trunk fully confiemed their suspicions and proved beyond any possible doubt that the hapless Marya Vasilievna was the perpetrator of all the petty thefts which had caused so much commotion during the past weeks.

"What a low, nasty thing she is! She even palmed the jam off on poor Feklusha to take attention away from herself and throw all the blame on the girl! Oh, the shameless woman! A little child, and she has no pity for her!" said Nanny in disgust and horror, completely forgetting her own role in the episode and how her own cruelty had forced poor Feklusha to give false testimony against herself.

One can picture the indignation of all the servants and of the household in general when the appalling truth was revealed and made known to all.

At first, in the heat of his anger, our father threatened to send for the police and have Marya Vasilievna put in prison. But in view of the fact that she was already a middle-aged, sickly woman who had lived in our house for so many years, he soon softened and decided merely to dismiss her and send her back to Petersburg.

One might think that Marya Vasilievna herself should have been satisfied with this sentence. She was such an expert needlewoman that she need never have feared going hungry in Petersburg. And what kind of position could she have held in our household after such an episode? All the rest of the servants had previously envied and disliked her for her pride and arrogance. She was aware of this and knew also how cruelly she would have to atone for her former grandeur.

Strange as it may seem, however, she was not only unhappy with my father's decision but, on the contrary, started begging his mercy. Some kind of feline attachment to our house came to the fore, perhaps, to her old familiar place in the world.

"I don't have long to live—I feel that I shall die soon," she said. "How can I go and live among strangers before I die?"

But Nanny, reminiscing with me many years later, when I was quite grown up, had an entirely different explanation. "It was just more than she could stand to leave us, because Filipp Matveyevich was staying on, and she knew that once she went away she would never see him again. If she, who lived her whole life as an honest woman, could do such a shabby thing in her old age, then she evidently loved him so much she couldn't stand it!"

As far as Filipp Matveyevich was concerned, he managed to come out of the water quite dry. It may be that he was really telling the truth when he maintained that in accepting presents from Marya Vasilievna he had no knowledge of where they had come from. In any case, since it was difficult to find a good gardener and our garden and vegetable plot could not be left to the whim of fate, it was decided to keep him on, at least for a time.

I do not know whether Nanny was right about the reasons impelling Marya Vasilievna to cling so stubbornly to her place in our house. Be that as it may, she went to my father on the day designated for her departure and threw herself sobbing at his feet.

"Better to let me stay on without pay, punish me like a serf—but please don't drive me away!"

My father was touched by this deep attachment to our household. But, on the other hand, he feared that if he forgave Marya Vasilievna, the rest of the servants would be demoralized. He was in great perplexity as to what to do when suddenly a plan came into his head.

"Listen here," he told her. "Stealing is a great sin, but I could have forgiven you anyway if your guilt consisted only in your thievery. But an innocent girl suffered because of what you did. Just think of it! On account of you Feklusha was subjected to such shame— a public whipping! For her sake, I cannot forgive you. If you truly wish to stay on with us, I can give my consent only on one condition: that you beg Feklusha's pardon and kiss her hand in the presence of all the servants. If you're willing to go that far, all right then—stay here!"

No one believed that Marya Vasilievna would consent to such a condition. How could she, a proud one like her, apologize publicly to a serf and kiss her hand? But suddenly, to everyone's astonishment, Marya Vasilievna agreed to do it.

Within an hour after her decision all the household was assembled in the entrance hall of our house to view the curious spectacle: Marya Vasilievna kissing Feklusha's hand. My father had demanded precisely that: that the event should take place with solemnity and in public. There was a large crowd. Everyone wanted to watch. The master and mistress were there too, and we children also asked permission to come.

I will never forget the scene which followed. Feklusha, embarrassed by the honor which had so unexpectedly fallen to her lot and fearful, perhaps, that Marya Vasilievna might avenge herself later for this compulsory humiliation, went up to the master and begged him to relieve her and Marya Vasilievna of the hand-kissing.

"I've forgiven her without it," she said, ready to cry.

But my father, who had tuned himself up to a high key and convinced himself that he was behaving in accordance with the precepts of strict justice, only shouted at her. "Get moving, you little fool, and don't stick your nose into other people's business! It's not for you that this is being done. If I had been guilty toward you—do you understand me? I myself, your master—then I, too, would have to kiss your hand. You can't understand that? Then hold your tongue and be quiet!"

The cowering Feklusha did not dare interpose any further objections. With her whole body shaking in terror, she went and stood in her place, awaiting her fate as if she had been the guilty one.

White as linen, Marya Vasilievna made her way through the crowd which parted before her. She walked mechanically, as if in her sleep. But her face was so rigid and angry that it was awful to look at her. Her lips were bloodless and convulsively pressed together. She came up very close to Feklusha. The words, "Forgive me!" tore from her lips in a kind of sickly scream. She grabbed Feklusha's hand and brought it to her lips so violently and with a look of such hatred that it seemed as though she wanted to bite it.

Suddenly a convulsion twisted her face and foam appeared at the corners of her mouth. With her whole body writhing, she fell on the ground and began screaming with piercing, inhuman shrieks.

It was discovered later that she had been subject to these nervous attacks—a form of epilepsy—even before that. But she had carefully concealed this fact from her masters, fearing that they would dismiss her if they found out. Those of the servants who knew about her disease kept their silence out of a feeling of solidarity.

I cannot convey the effect her seizure had on those

present. It goes without saying that we children were hastily taken away. We were so terrified that we were close to hysterics ourselves. But even more vividly I remember the sudden shift which took place in the mood of all our household servants. Up to that time they had behaved toward Marya Vasilievna with anger and hatred. Her act seemed so vile and low that each one derived a certain pleasure from showing her his contempt, from spiting her in some way.

But now all that was changed suddenly. She had unexpectedly appeared in the role of suffering victim, and popular sympathy shifted over to her side. Among the servants there was even a repressed protest against my father for the excessive severity of his punishment.

"Of course she was wrong to do what she did," the housemaids would say in undertones when they gathered in our nursery to confer with Nanny, as was their habit after every important event. "Well all right then, so the General could have given her a good tonguelashing, the mistress could have punished her herself, the way it's done in other houses. That doesn't hurt so much, you can bear it. But now, all of a sudden, see what they thought up! To go and kiss the hand of such a little cricket, such a snotnose as Feklusha, right in front of everybody! Who could stand such an insult!"

Marya Vasilievna did not regain consciousness for a long time. Her seizures recurred again and again over an interval of several hours. She would blink, become conscious for a mement and then suddenly start thrashing around and screaming again. The doctor had to be called from town.

With each passing minute, sympathy for the patient increased and indignation against the masters grew. I remember my mother coming into the nursery in the middle of the day. Seeing Nanny brewing tea with a good deal of fuss and concern at that unusual hour, she asked quite innocently, "For whom are you doing this, Nanny?"

"For Marya Vasilievna, naturally! What do you think—is it your opinion that she, a sick woman, should be left without tea? We servants, we still have a Christian heart!" Nanny replied, in such a coarse and challenging tone of voice that my mother grew quite embarrassed and hurried away.

And yet a few hours earlier, that very same Nanny, if she had been given her way, would have been capable of beating Marya Vasilievna half to death.

The seamstress recovered within a few days, to my parents' great joy. She took up her life in the house just as before. No one mentioned what had taken place. I believe that even among the servants there was no one who would have reproached her for the past.

But as for me, from that day on I felt a strange pity for her, mixed with an instinctive horror. I no longer ran to her room as I used to do. If I met her in the hall I couldn't keep from pressing myself against the wall, and I tried not to look at her. I kept imagining that she would fall on the floor right then and there and start thrashing and screaming.

Marya Vasilievna must have been aware of my alienation from her, and she tried to win back my old affection by various means. I remember that almost every day she would think up different little surprises for me: now she would bring me colored scraps of cloth, now she would sew a new dress for my doll. But none of this helped. The feeling of secret terror would not pass, and I ran away the moment I found myself alone with her. And soon after that, I came under the supervision of my new governess, who put a stop to all my friendly relations with the servants.

But I vividly recall the following scene. I was already seven or eight years old. One evening, the night before some holiday—the Annunciation, perhaps—I was running down the hall past Marya Vasilievna's room. Suddenly she looked out and called to me.

"Young lady, young lady! Come in and see me. Look what a lovely lark I baked for you out of dough!"

It was half dark in the long hall, and no one was there but Marya Vasilievna and myself. Looking at her white face with its great, dark eyes, I suddenly felt an eerie sensation. Instead of answering her, I dashed away headlong.

She called after me. "What is it, young lady? I can see that you don't like me at all any more. I disgust you!"

It wasn't so much her words as the tone of voice in which she said them that shook me. I didn't stop, but kept on running. But then, on returning to the classroom and calming down after my fright, I couldn't forget the sound of that voice—hollow, despondent.

I was not myself all evening. No matter how I tried to suppress the unpleasant gnawing sensation inside of me by playing, by prankishness, I couldn't make the feeling go away. The thought of Marya Vasilievna wouldn't leave my mind. And, as always happens with a person one hurts, she suddenly seemed terribly nice to me, and I began to feel drawn to her.

I couldn't bring myself to tell my governess what had happened. Children are always embarrassed to talk about their feelings. Moreover, since we were forbidden to fraternize with the servants, I knew that the governess would in all likelihood praise me for my behavior. And yet I felt with every instinct that there was nothing praiseworthy about it.

After evening tea, when it was time for me to go to bed, I decided to drop in to see Marya Vasilievna instead of going straight to my room. This was a kind of sacrifice on my part, for it meant running alone down a long, deserted, and by now quite dark hall which I always feared and avoided in the evening. But now a desperate bravery came to the fore. I ran without stopping to take a breath. Puffing and panting, I tore into her room like a hurricane.

Marya Vasilievna had already had her supper. Because of

the holiday, she wasn't working but sitting at the table, covered with a clean white cloth, and reading some religious book. The lamp glimmered in front of the icons. After the frightening dark hall, the little room seemed uncommonly light and cozy, and Marya Vasilievna herself so kind and good!

"I came to ask you to forgive me dear dear Marya Vasilievna!" I said in one breath. Before I could finish, she had already grabbed me and started covering me with kisses. She kissed me so violently and for such a long time that I felt the eerie sensation once more. I was already trying to figure out how to get out of her grasp without offending her again, when a cruel attack of coughing forced her to release me from her embrace at last.

This dreadful cough tormented her more and more. "I barked like a dog all night," she would say of herself, with a kind of sullen irony.

With each day that passed she grew paler and more withdrawn, but she stubbornly resisted all my mother's suggestions that she consult a doctor. She even showed an angry irritation when anyone mentioned her illness.

In this way, she dragged out another two or three years. She was on her feet almost to the end. She went to bed only a few days before she died; and her final hours, they said, were horribly painful.

My father ordered a very opulent funeral (by village standards) to be arranged for her. Not only all the servants, but all our family attended it as well, even the master himself. Feklusha, too, walked behind the coffin and sobbed bitterly. The only one missing was Filipp Matveyevich. He did not wait for her to die. He had left us a few months earlier for another and better-paying job, somewhere in the vicinity of Dinaburg.

CHAPTER THREE

Metamorphosis

With our move to the country, our household underwent a radical change. All at once my parents' life, so gay and lighthearted until then, took a more somber turn. My father had paid us little notice up to that time, for he considered bringing up children to be a woman's and not a man's affair. He was more concerned with Anyuta than with the other children, for she was the eldest, and she was a very amusing child. He loved to find occasions for indulging her. He sometimes took her sledding in winter, and liked to boast about her to our guests. When her naughtiness went beyond all bounds and exasperated the servants past endurance, they sometimes went to Father to complain about her. But he would turn the whole thing into a joke, and Anyuta understood perfectly that even though he sometimes put a stern expression on his face for appearance's sake, he was inwardly ready to laugh her naughtiness away.

As for us younger children, our father's contact when he saw us was limited to asking Nanny whether we were well and giving our cheeks an affectionate pinch to make sure they were properly plump. Sometimes he would lift us in his arms and toss us up in the air. On state occasions, when he was going out to make some official appearance, decked out in full-dress uniform complete down to the decorations and the stars on his chest, we would be called into the drawing room to "come look at Papasha all dressed up." This spectacle gave us uncommon pleasure. We would skip all around him clapping our hands with glee at the sight of his gleaming epaulettes and military orders.

But when we moved to the country the mellow relationship existing between Father and us changed suddenly. As not infrequently happens in Russian families, Father all at once made the unlooked-for discovery that his children were far from being the exemplary, beautifully brought up children he had assumed they were.

It all started, I think, from the time my sister and I ran away from the house, got lost and disappeared for an entire day. When they found us toward evening we had managed to stuff ourselves with inedible berries and were sick for the next few days. This episode revealed the fact that our supervision was lax in the extreme. Other disclosures followed fast upon the first; revelation succeeded revelation.

Up to that time, everyone had staunchly believed that my sister was virtually a phenomenal child, intelligent and cultivated beyond her years. Now, however, it turned out that she was not only spoiled through and through, but also appallingly ignorant for a twelve-year-old, and didn't even know how to spell.

And there was something even worse than that. A bad thing came to light about our French governess, a thing that must not even be mentioned in front of us children.

I have a cloudy recollection of those melancholy days after our escapade as a kind of painful domestic calamity. In the nursery there was shouting, crying, hubbub all the livelong day. Everybody quarreled with everybody else, and everybody was punished—the guilty and the innocent alike. Papa was enraged, Mama wept, Nanny bellowed, the French governess wrung her hands and packed her bags.

My sister and I grew quieter and more subdued and didn't dare open our mouths; for by now, each of the grownups was venting his irritation on us, and the most trivial prank would get us into deep trouble. All the same, we followed the adults' bickering with curiosity and even with a certain childish gloating, and we waited to see how all this was going to come out.

Father, no lover of half measures, resolved upon the radical transformation of our entire system of education. The French governess was dismissed. Nanny was removed from the nursery and assigned to look after the household linens. And two new personages were taken into our house: a Polish tutor, Yosif Malevich,[1] and an English governess, Miss Margaret Smith.[2]

The tutor turned out to be a quiet, knowledgeable man who taught superbly, although his actual influence on my education was slight. The governess, on the other hand, introduced a totally new element into our household. Although she had been brought up in Russia and spoke Russian well, she had retained intact all the typical characteristics of the Anglo-Saxon race: fixity of purpose, self-control, and the ability to carry things through to their conclusion. These qualities gave her an enormous advantage over the rest of the household, which was distinguished by quite antithetical traits, and they explain the influence she acquired in our home.

When she came to us, she directed all her efforts toward reorganizing our children's wing into a proper English *nursery** in which she could rear model English misses. And

---

*The word is given in English in the Russian text [Tr.]

goodness only knows how difficult it was to cultivate a seed-plot of English misses in a Russian manorial home implanted for ages and generations with gentry customs, slovenliness and laxity. Thanks to her remarkable persistence, however, she succeeded to a degree in achieving her ends.

It is true that Miss Smith never quite learned how to cope with my sister, who had been accustomed to complete freedom up to that point. The next year-and-a-half or two passed in continual clashes and confrontations between them. Finally, when Anyuta was fifteen years old, she stopped obeying the governess altogether.

This de facto emancipation from Miss Smith's supervision took the form of having her bed moved out of the nursery and into the room next door to Mama's bedroom. From that day forward my sister began to regard herself as a grownup young lady, and the governess missed no opportunity to demonstrate tangibly that Anyuta's behavior was not of her making, that she washed her hands of it.

But in compensation she now focused her entire attention on me with even greater intensity than before, isolated me from the other members of the family and barricaded me from my elder sister's influence as if from an infectious disease. This urge toward separatism was abetted by the dimensions and arrangement of our country house, which three or four families might easily have occupied simultaneously while remaining total strangers to one another.

The whole lower floor, except for a few rooms set aside for the servants or for casual guests, was given over to Miss Smith and to me. The upper floor, including the front reception rooms, belonged to Mama and Anyuta. Fedya and the tutor were housed in another wing, and Papa's study comprised the foundation of a three-story tower situated quite apart from the rest of the house. In this way, the different elements which made up our family kept each to its own independent domicile, and each could maintain its separate regime without impinging upon the others—coming together only at dinner and for evening tea.

CHAPTER FOUR

Palibino[1]

The locale of our[2] estate was very wild but more pictorial than most of the regions of Russia's central zone. Vitebsk province is renowned for its vast conifer forests and its numerous large and beautiful lakes. The last spurs of the Valdai Hills pass through parts of it, so that there are no such immense plains as in central Russia. On the contrary, the entire landscape has a rolling, undulating character.

As in most of Russia, there is little rock to be found. And yet one sometimes comes unexpectedly upon an enormous granite boulder in the middle of some flat field or water-meadow with grass growing as tall as a man. This boulder stands out so strangely from the lush greenery surrounding it, is so at variance with the gently rounded contours of the countryside as a whole, that one cannot help wondering how it landed there.

A thought enters one's head unbidden: might it perhaps be a monument left behind by some unknown or even supernatural beings? And in fact, geologists maintain that the boulder is an alien visitor carried here from far off, that it is indeed a curious monument—but not one created by an extinct people nor by fabled gnomes. Rather, it is a relic of that great Ice Age when gigantic cliffs were torn away from Finland's shores like tiny grains of sand, borne over vast distances under the shattering pressure of the slowly advancing ice.

One side of our estate abutted upon wooded land almost to its very edge. At first this wood was kept cleared and looking like a park, but little by little it had grown denser and more impassable, until at last it merged with the vast State forest. This latter extended over an area of hundreds of versts. No axe had ever sounded in it in the memory of man except clandestinely, perhaps, in the hands of some poaching peasant come to make off with government wood by dark of night.

Strange legends about this forest circulated among the peasants, and it was hard to determine where truth left off and myth began. Like all Russian forests, it was of course, inhabited by a variety of evil spirits such as wood-demons and river-nymphs. Hardly a soul doubted the existence of these creatures but nobody, to tell the truth, had seen them with their own eyes except for Grunya the village simpleton or the old sorcerer Fedot.

On the other hand, there were many more people who could tell a tale of encountering this or that wicked man. Rumor had it that whole dens of brigands, horse-thieves and Army deserters were hidden in the depths of the woods; and if the local constable or district police officer should ever take it into his head to investigate the goings-on in the forest at night, he would find himself in trouble. As for wolves, lynx and bears, there was almost no one among the locals who hadn't had at least one occasion to find out from personal experience that these creatures did roam the forest, beyond doubt.

It should be added that the bears maintained a peaceful enough coexistence with the local peasantry. Occasionally in early spring or late fall one might hear that a bear had killed a peasant's cow or horse. As a rule, however, the bear was content to filch a few sheaves of oats from his neighbors' fields or honey from their hives. Very rarely, a rumor would suddenly start up that a bear had mauled a peasant, but even then, it always came out later that it was the peasant himself who was at fault for having provoked the innocent bear.

Many people harbored an almost superstitious dread of the forest. If, as sometimes happened, a peasant woman in one of the outlying villages should miss her child of an evening, the first thought that crossed her mind was that he had lost his way in the forest, and then she would begin to wail as though the child were dead.

Not a single maid in our household dared to venture into the forest alone, without a companion. But in groups, particularly in the company of the young menservants, they ran to the woods very willingly, of course. Our intrepid English governess, who had a passion for constitutionals and long walks, was very sceptical at first about all the stories people tried to frighten her with as soon as she arrived at Palibino. She decided to take regular strolls in the forest, in disregard of the fearful women's horror stories.

One day in fall, however, when she and her charges were out alone no more than an hour away from the house, she suddenly heard a crackling sound in the woods, and then she caught sight of a huge she-bear crossing the path with her two cubs, perhaps fifty paces away. After that she was forced to concede that not everything in the forest stories was exaggerated. And she too stopped venturing further than the edge of the woods, unless she was accompanied by one of the menservants.

And yet, it was not only frightening things that came out of the forest. It had inexhaustible stores of riches of every sort. There were numberless varieties of game: hares, black grouse, hazel grouse, partridge. All a hunter had to do was walk in and shoot. Even the clumsiest, with only an old flintlock for a rifle, could count on getting a bagful. And in summer there was no end to the different kinds of berries. First to appear would be the wild strawberries which, it is true, ripen a bit later in the woods than in the fields, but then are much juicier and more fragrant. Almost before the strawberries were gone the bilberries would make their appearance, and then the stone-fruit, the raspberries, the cranberries.

And then, before you realized it, the nuts would be ripening, and after that the mushroom harvest would begin. Even in summer one might find a good many brown-caps and orange-caps, but autumn was the real season for the "milkies," the "rusties," and the prized "whites."

In all the villages round about, a kind of frenzy would come over the women, the young girls and even the children. They couldn't be pried out of the forest by force. They would set off in a throng at daybreak with their bowls, their woven and bast baskets, and not come home until late in the evening.

And how greedy they were! No matter how big a crop they managed to reap from the forest that day, it was never enough. The next day they'd be lured back at the first gleam of light. All their thoughts were focused on gathering mushrooms; for the sake of mushrooms all of them were ready to neglect their work, both at home and in the fields.

Expeditions into the forest were undertaken in our house as well, sometimes in summer at strawberry time, sometimes in fall during the mushroom season. The entire household took part with the sole exception of the master and the mistress, neither of whom cared for rustic pleasures of that sort.

All the arrangements were made the night before. With the first rays of the rising sun the next morning, two or three wagons would already be approaching the porch. Inside the house a merry, festive bustle was beginning. The maids scurried back and forth, rushed about carrying supplies and stowing them inside the wagons: plates, a samovar, various provisions such as tea, sugar, bowls filled with the meat turnovers and cheese tarts called *pirozhki* and *vatrushki,* baked by the cook on the previous day. On the top of the heap they threw the empty baskets and containers intended for the mushroom gathering.

The children, roused from their beds at this unaccustomed hour, with sleepy faces which had just been rubbed bright pink by a wet sponge, were also running about. In their glee they didn't know what to do first, grabbed at everything, interfered with everybody and invariably managed to get under everyone's feet.

The kennel dogs were no less interested in the forthcoming excursion. They had been in a state of nervous excitement since early morning and had been running under people's legs, peering into their faces, yawning long and loud. Finally, exhausted with excitement, they sprawled in the courtyard in front of the porch, but their entire posture expressed strained expectation: they watched the whole to-do with uneasy stares, ready to jump up and tear away at the first signal. Every fiber of their canine being was now concentrated on a single aim—not to let the masters go off and leave them behind!

Now at long last the preparations were finished. Helter-skelter into the carts piled the governess, the tutor, the children, a dozen or so housemaids, the gardener, two or three of the menservants, and also perhaps five or so of the servants' children. The whole servant body was wrought up with excitement. Everyone wanted to take part in the festive outing. At the very last moment, when the carts were on the point of moving off, the scullery maid's little daughter, five-year-old Aksyutka, came running up and raised such a racket when she saw her mother going away and leaving her behind, that she too had to be put into the cart.

The first stop was to be at the forester's lodge, about ten versts distant from our estate. The carts jogged down the marshy forest path at an easy trot. Only the lead cart had a real coachman sitting on the box: the rest were driven by volunteers who kept grabbing the reins out of each other's hands and clumsily jerking the horses now to the right, now to the left.

A sudden jolt. One of the carts had driven over a thick root. All the passengers were tossed up in the air. Little Aksyutka barely escaped falling out altogether. Someone managed to grab her by her dress collar as one picks up a puppy by the scruff of its neck. An ominous clink of broken glass was heard at the bottom of the cart.

The forest was becoming ever denser and deeper. Wherever you looked were fir trees, somber, tall, with rough dark brown trunks erect as giant church tapers. Only along the edges of the road crept smaller shrubbery—hazelnuts, elderberry bushes, and mainly thickets of alder. Here and there a trembling aspen leaf, reddened by autumn, flashed past, or a picturesque rowan tree, heavy with berry clusters, blazed bright.

Suddenly loud, frightened cries were heard in one of the carts. The improvised driver had brushed his cap against a young birch branch bending low over the path and still wet with dew. The branch swayed, and with a sweep it lashed everyone sitting in the cart and showered them with tiny, fragrant sprinkles. There was no end to the laughter, the jokes, the witticisms.

Now the forester's lodge came into view. His hut was roofed with shingles and looked much cosier and cleaner than the usual Belorussian peasant's hut. It stood in a little clearing and (a rare luxury for peasants in those parts) was surrounded by a small kitchen garden where poppies made a flash of color among the cabbage heads and two or three bright yellow sunflowers glowed. The few apple trees rising above the vegetables, heavy with rosy fruit, were their owner's special pride, for he himself had transplanted them from the woods where they were growing wild, then cultivated them. And now his apples rivaled those of any of the landowners in the vicinity.

The forester was getting on toward seventy. His beard was very long and quite white, but he looked hale and hardy still, very staid and sedate. In height and build he was bigger than most small-boned Belorussians, and his face seemed to reflect the forest in its clear and majestic calm. All his children were grown and settled in life: his daughters were married, his sons were apprenticed elsewhere in some craft. And now he lived alone with his old wife and also with a foster child, a lad of about fifteen, whom he had taken into his home in his old age.

Catching sight of the masters' procession from a distance, the old woman hurried to get the samovar started. As the carts drove up to the porch, both she and the old man came out to greet them, bowing from the waist and begging their dear guests not to disdain their tea.

The inside of the hut was also clean and tidy, although the air was stuffy and saturated with the smell of incense and lamp oil. As a precaution against winter's frosts, the windows were minuscule and hard to open. After the fresh spaciousness of the woods, it seemed impossible even to draw a breath for the first few moments.

But there were so many interesting things inside the hut that we children soon got used to the stuffy air and began looking around curiously. Fir branches were spread over the clay floor. Benches were arranged along every wall, and a tame jackdaw with clipped wings was hopping on them, not a bit disconcerted by the big black tomcat who was evidently its good friend. The cat sat on its hind paws washing its muzzle with one of its front ones; with a show of indifference, it watched the visitors through half-closed lids.

In the front corner stood a large wooden table covered

by a white cloth with an embroidered hem. Above it towered a huge icon-case containing extremely ugly and evidently very ancient icons.

It was said that the forester was an Old Believer[3] and that this was why he lived so tidily and comfortably, for it is well known that Old Believers never frequent taverns and maintain great cleanliness in their dwellings. It was also said that every year the forester had to buy off both the district police officer and the village priest at no small expense. In exchange for this, the latter did not interfere with the forester's religious beliefs and did not force him to attend Orthodox church services nor check up on his attendance at the Old Believers' chapel.

And still more things were said about him; for instance, that he would never touch a bite of food in an Orthodox household, and that he kept a set of special dishes for Orthodox guests. Even though these guests might be gentry folk, still he would never, under any circumstances, serve them any food from his own dish, for that would mean defiling the dish just as though a dog or some other unclean animal had eaten from it.

We children were very eager to know whether it was true that Uncle Yakov, as we called him, felt squeamish toward us, but we didn't dare to ask him. We liked Uncle Yakov very much. We considered it a great pleasure to be in his house. When he made an occasional visit to us at Palibino he always brought us some gift which was much more to our liking than the most expensive toy. Once, for example, he brought along a young elk which lived for a long time behind the fence in the Palibino park, although it never grew completely tame.

The huge copper samovar was puffing away on the table where a variety of unusual foods was set out: *varenets,* a thick baked sour-milk dish*, *lepyoshki,* soft flat cakes, made

---

*Kovalevskaya's note [Tr.]

with poppyseed, cucumbers in honey, all kinds of goodies we children never got except at Uncle Yakov's. He zealously plied his guests with food but touched nothing himself ("That means it's true," we thought to ourselves, "—he does feel squeamish toward us"), and he carried on a sedate, unhurried conversation with our tutor. Quite a few incomprehensible Belorussian expressions were mixed in with his speech, but we loved to listen to him all the same. He knew so much about the forest, the animals in it, and what every animal thinks.

By now it was about 6 A.M. How strange! On ordinary days we'd be asleep in bed at this hour, but today, just think of all the things that had happened already! No more time to dawdle. Our whole company scattered through the forest, calling back and forth and hallooing from time to time, so as not to separate too widely and get lost.

Which one of us would gather the most mushrooms? This was the question now agitating everyone. Each burned with ambition. It seemed to me at that moment that nothing in the world was more important than filling up my basket as fast as possible. "God!" I thought (involuntarily putting a great deal of fervor into my prayer), "please send me lots and lots of mushrooms!"

Catching a glimpse of an orange or a blackish-brown cap from a distance, I raced to the spot as fast as my legs would carry me, so that no one would snatch my find away. But there were so many disappointments! Now I mistook a dry leaf for a mushroom; then I suddenly spied the firm beige cap of a "white" rising shyly from its bed of moss. I seized it, thrilled. But lo and behold, from underneath it was not solid white, but deeply furrowed gills. So there it was—nothing but an ordinary baby toadstool which from above had assumed the deceitful look of a "white!"

But the most upsetting thing of all to me was walking past a place in complete obliviousness while sharp-eyed Feklusha would snatch up a beautiful mushroom practically from under my nose. That horrid Feklusha—she could sniff out by instinct where the mushrooms were, she simply dug them up from under the earth. Her basket was already heaped to the brim. And what's more, almost all of her mushrooms were "whites" and "rusties." Some orange-caps, perhaps—just a few. But as for the "foxies," the "butters" and the "bitters"—she didn't bother with those at all.

And such lovely mushrooms she had! Choice, every last one: small, clean, pretty—you could eat them raw! Whereas my basket was still only half full, and so many of those were big, shriveled old caps that I was ashamed to show them to anybody.

At three o'clock there was another break. The coachman lit a campfire in the clearing where the unharnessed horses were grazing. A lackey ran to the nearby stream to fill the carafes with water. The maids spread a tablecloth on the grass, started the samovar going, arranged the dishes and plates. The masters sat down in a separate group, while the servants respectfully placed themselves a slight distance away.

But this division lasted no more than the first fifteen minutes. Today was such a special day that class distinctions seemed not to exist. Everybody was possessed by the same all-consuming interest: mushrooms. Therefore the company soon intermingled again. They all wanted to brag about their loot and to see what the rest had turned up. And besides, they had so many things to tell each other now. Each had had his own adventures: one had startled a hare, another discovered a badger's burrow, a third almost stepped on a snake.

After eating and resting for a while, they went back for more mushrooms. But the first ardor had evaporated by then. The tired feet dragged along laboriously. The big basket, even though it now held only a few mushrooms, had suddenly grown so heavy that it weighed on the arm.

The inflamed eyes refused to function. Either they imagined mushrooms in places where there were none, or they passed right over a genuine mushroom without seeing it at all.

By now I no longer cared whether I would have a basketful or not. But then I became far more sensitive to other forest impressions. The sun was already close to setting. Its slanting rays glided between the naked trunks and tinted them the color of brick. The little forest lake, set into perfectly flat banks, was so unnaturally calm and motionless that it seemed enchanted. The water in it was very dark, almost black except for one bright crimson spot that glowed like a blood-red stain.

Time to go home. The whole company congregated at the wagons once again. During the day they had all been so engrossed, each in his own affairs, that no one paid attention to the others. But now they looked each other over and burst out laughing. Goodness only knows what they looked like! In the course of this day spent in the open air, they had all felt the effects of the sun. All the faces were windburned and flaming. Their hair was disheveled, their clothes were in indescribable disorder. Both mistresses and maids had put on their oldest dresses, of course (the ones not worth saving), for their excursion into the forest—in the morning all that hadn't seemed to matter.

But now they were a ludicrous sight. One girl had lost her shoes in the woods. On another, what had once been a skirt now hung on her in tatters. The headgear was the most fantastic of all. One of the girls had thrust a big cluster of bright red rowanberries into her unkempt black braids; another had fashioned herself a kind of helmet from fern foliage; a third had pushed a stick through a monstrous death's-cap mushroom and was holding it over herself like an umbrella.

I had wound a flexible branch of forest hops around myself. Its yellow-green cones had become entangled with my tousled brown hair, which tumbled down to my shoulders and gave me the look of a little Bacchante. My cheeks were flaming, my eyes sparkled.

"Hail to Her Majesty, Queen of the Gypsies!" said my brother Fedya, bending his knee before me in mockery.

The governess also had to admit with a sigh that I really did look more like a gypsy girl than a well-brought-up young lady. But if the governess had only known how much I would have given at that moment to be transformed into a real gypsy! That day in the forest had awakened in me so many wild, nomadic instincts. I wanted never to have to go home again; I wanted to spend the rest of my life in that lovely, marvelous forest. Such dreams, such fantasies of distant journeys and fabulous adventures teemed inside my head. . . .

The trip home took place in deep silence. There were no more merry shouts or bursts of laughter as in the morning. Everyone was tired and subdued. A strange, almost solemn mood descended on us all. Some of the girls stuck up a song, so soft and doleful that my heart contracted with the strange, inexplicable longing that often came over me after moments of intense excitement. But this longing had so much of its own peculiar charm that I would not have exchanged it for clamorous happiness.

After I was at home and lying in my bed, in spite of my fatigue I could not fall asleep for a long time. In a fevered state midway between sleep and waking, I could still see the forest before my eyes. If anything, I saw it more clearly, I grasped its general outlines better, and at the same time was more aware of its details than I had been during the day when I was actually there.

Many ephemeral impressions, which at the time merely glided past me without coming directly into consciousness, now returned, vividly and insistently. Here was an enormous anthill looming up out of the darkness; each fir needle on it projected in such graphic relief that I felt as if I could pick it up. The ants dragged their white eggs behind them busily and quickly until, suddenly, all of them vanished somewhere together with the anthill. And in their place appeared a soft white lump like a big clump of snow.

Now I could see that the whole thing was made of tiny gossamer spiderwebs, and in its very center was a little dark speck. I wanted to grasp that clump in my hand, but the thought had no time to take form before the black speck in the center spun into rapid motion, and black dots spilled out of it in all directions like radii from a center to a circumference. Only they weren't dots, they were tiny black baby spiders, all of which ran and scurried about.

That morning, in fact, I really had discovered a strange clump like this one, but I hardly noticed it then. And now I could see it so vividly, as if it were real. And tired as I was, I twisted and turned for a long time in my bed and couldn't shake off these intrusive visions, until finally I fell into a heavy, leaden sleep.

The forest, which played so large a role in my childhood memories, adjoined one side of our estate. On the other lay a garden sloping down to a lake, with fields and meadows extending beyond that. Here and there amid the green expanse emerged little villages, looking more like animal burrows than human habitations.

The soil of Vitebsk province is not nearly so fertile as that of the black earth belt of Russia and the Ukraine. Belorussian peasants are noted for their poverty. Emperor Nicholas, driving through this region once, aptly called Belorussia "the poor beauty" in contrast to Tambov province, which he dubbed "the rich merchant's wife."

In the middle of this wild and thinly populated area the Palibino manor, with its massive stone walls, its fanciful architecture, its terraces framed in summer with garlands of roses, its huge orangeries and hothouses, stood out in strange contrast. In summer there was still some liveliness in those parts, but in winter it all came to a standstill and grew desolate.

Snow swept over all the garden paths and piled in great drifts up to the very verge of the forest. Wherever you looked out of a window, there was nothing but white and lifeless plain. No one was to be seen traveling the high road for hours on end. Once in a while some peasant sledge would drag along, hitched up to a skinny old nag all white with hoarfrost; then, for a long while, no sign of life or movement, as though everything were numb.

Sometimes, at night, wolves would come quite close to the house. We would all be sitting at evening tea. The crystal chandelier was lighted in the large adjoining salon, and the candleflames merrily reflected back and multiplied in the big wall mirrors. Expensive damask-upholstered furniture was arranged along the walls. The big cut leaves of palms and other greenhouse plants made fanciful patterns in front of the windows. Books and foreign journals were strewn about on tables.

Teatime was over, but the children had not yet been sent off to bed. Father smoked his pipe and laid out a game of patience. Mama sat at the piano playing a Beethoven sonata or a Schumann romance. Anyuta was pacing back and forth through the salon, carried far away in her imagination, far from reality. She was seeing herself in a brilliant company as queen of the ball.

Suddenly Father's footman, Ilya, appeared at the door. He said nothing, but he shifted his weight from one leg to the other, his inveterate habit when he was getting ready to convey some special piece of news.

"What is it, Ilya?" Father asked at last.

"Nothing, Your Excellency," Ilya replied with an odd kind of smile. "I only came to report that there are lots of wolves gathered on our lake. Wouldn't the masters be interested to go and listen to them howl?"

At this news, we children of course became wildly excited and begged to be allowed to go outside on the porch. After expressing some apprehension lest we catch cold, Father gave in to our pleas at last. We were dressed in warm fur coats, our heads were wrapped in down shawls, and we went out on the terrace accompanied by Ilya.

It was a glorious winter night. The frost was so sharp that it took our breath away. Although the moon was not out, there was light from the masses of snow and the myriads of stars studded like big nails across the whole expanse of sky. I thought that I had never seen such brilliant stars before that night. They seemed to be throwing their rays to each other like balls, and each one shimmered marvelously, now flashing in a spurt of brilliance, then dimming down for a moment.

Snow, masses of snow everywhere, whole mountains of snow covering and leveling everything. The terrace steps could not be seen at all. One wasn't even aware that the terrace was higher than the garden surrounding it. It was all transfigured now into one even white plain merging imperceptibly with the frozen lake.

But even more extraordinary was the hush in the air, a deep, unbroken silence. We children stood on the porch for a few minutes, but we still heard nothing. We began to feel impatient, "Where are the wolves, then?" we asked.

"Gone and quieted down, they have, on purpose-like," Ilya answered, annoyed. "But just you wait a bit and maybe they'll start up again in a couple of minutes."

And indeed, a long-drawn out, rolling howl suddenly resounded and was answered immediately by several other voices. And now from the lake came a chorus so strange, so inhumanly dismal, that our hearts stood still.

"There—there they go, our friends!" Ilya exclaimed in triumph. "They've started up with their tunes. And what made them go and take such a liking to our lake? I can't figure it out. They go out there the minute night comes down—dozens of them."

He turned suddenly to our general favorite, the big Newfoundland dog, who had jumped out on the terrace together with us. "How about that, friend Polkan? Maybe you feel like going over to them? Maybe you want to join in with them and get yourself a little taste of wolves' teeth?"

But the concert of the wolves had a terrifying effect upon the dog. Normally so chipper and always ready for a fight, he now huddled close to us with his tail between his legs. His whole look expressed overwhelming fear. We children also began to be disturbed by the strange and savage chorus. We felt a nervous trembling going through us, and we hurried back inside to the snug, warm room.

CHAPTER FIVE

Miss Smith

The clock on the classroom wall struck seven. Those seven repeated strokes reached me through sleep and gave rise to the sad certainty that now, this minute, Dunyasha, the chambermaid, would be coming in to wake me up. But sleeping was still so sweet that I tried to convince myself that those seven repulsive strokes of the clock were only in my imagination.

I turned over on the other side and pulled the blanket tighter around myself. I hurried to extract every bit of that sweet, brief bliss afforded by the last moments of sleep, the finish to which, I knew, was coming right away.

And indeed, there was the creaking of the door, there were Dunyasha's heavy footsteps as she came into the room carrying an armload of firewood. Then a series of familiar sounds repeated every morning: the heavy banging of wood on the floor, the striking of matches, the crackle of kindling, the rush and rustle of flame.

All these familiar noises reached my ears through sleep and intensified the sensations of delicious luxuriating and reluctance to part with my warm bed. If only I could sleep one minute more, just one little minute!

But now the rustle of flame in the stove grew louder and more even and turned into a measured, steady drone. "Time to get up, young lady!" sounded directly over my ear, and Dunyasha pulled the blanket off me with a merciless hand.

Outside it was just turning light. The first faint rays of a cold winter morning, fusing with the yellowish light of the stearine candle, gave a kind of deathly, unnatural look to everything. Is anything on earth less pleasant than getting up by candlelight?

I squatted on my heels in bed and started to dress slowly, mechanically, but my eyes closed again of their own accord, and the lifted hand clutching the stocking stiffened in that position. Behind the governess's screen I could already hear the splash of water, the sounds of snorting and vigorous rubbing and drying.

"Don't dawdle, Sonya!" the governess said in English, and with a threat in her voice. "If you are not ready in a quarter of an hour, you will bear the ticket "LAZY" on your back during luncheon."

One did not make light of that threat. Corporal punishment had been banished from our educational regime, but the governess had contrived to replace it with other methods of instilling fear. Whenever I did something wrong, she would pin a piece of paper to my back with my sin blazoned on it in large letters, and I had to come to the luncheon table thus adorned.

I was deathly afraid of this punishment. The governess's threat, therefore, possessed the power to drive my sleepiness away in an instant. I jumped straight out of bed. The maid was already waiting for me at the wash basin, with a raised pitcher in one hand and a big, shaggy towel in the other. Every morning cold water was poured over me, in the English style. One second of biting, breath-stopping cold, then a sensation like boiling water coursing through my veins, and finally, what remained was an astonishingly pleasant feeling of springiness and well being.

By now it was quite light. We went into the dining room. The samovar puffed away on the table, the wood crackled in the stove, and a brilliant flame reflected back and multiplied in the big frosted windowpanes. Not a trace of drowsiness was left in me. On the contrary, there was such a good, such an inexplicably joyous feeling in my heart—how I yearned for some noise, some laughter, some gaiety! If only I had a friend, another child my own age with whom I could be silly, make a fuss—a child with an excess of healthy, young life bubbling over in her just as it was bubbling over in me. . . .

But I had no friend like that. I sat and drank my tea alone with Miss Smith, since the rest of the family—including my brother and my sister—got up much later than I.

I wished so hard for something to feel glad about and to laugh over that I even made some feeble efforts to play with Miss Smith. Unfortunately, she was out of sorts, a frequent morning occurrence because she suffered from a disorder of the liver. She therefore regarded it as her duty to bring my inappropriate surge of gaiety within bounds by observing that now was the time for studying and not for laughing.

My day began with a music lesson. It was always extremely chilly in the big salon upstairs where the piano stood, so my fingers would grow numb and swollen, and the fingernails would stand out on them like blue spots. An hour and a half of scales and exercises, accompanied by the monotonous striking of the stick with which the governess beat time, subdued once and for all the feeling of exuberance with which I had begun the day.

The music lesson was followed by other lessons. During the time when my sister was studying too, I took great pleasure in my own lessons. And I was still so little then that my education was regarded very lightly. But I used to wheedle permission to come to Sister's class, and I would listen so attentively that the next time when, as often happened, she, a great big fourteen-year-old, didn't know her homework assignment, I, a seven-year-old chit of a girl, would remember the answers and prompt her triumphantly. I took great delight in that.

But now that my sister had given up her studies and assumed her adult rights, my lessons had lost half their charm for me. I studied diligently enough, it's true. But how zealously I would have studied if only I had a friend!

Luncheon at noon. Miss Smith swallowed her last bite and walked over to the window to check on the weather. I followed her movements with beating heart, for this question was very important to me. If the thermometer read higher than ten below zero and if the wind was not strong, I was faced with a dismal hour and a half together with my governess, walking back and forth along the path, which had been cleared of snow. But if, to my joy, there was a severe frost or high winds, she would set out alone on that walk, absolutely essential by her lights, while I was sent upstairs to get my exercise by playing ball in the drawing room.

I didn't particularly care for playing ball. I had already reached the age of twelve. I considered myself quite grown up, and I was even a little insulted that Miss Smith still thought me capable of getting fun out of such a childish pastime as ballplaying. All the same, I was very glad to hear this command because it heralded an hour and a half of freedom for me.

The upper floor was Mama's and Anyuta's special property, but at the moment they were both in their own rooms. The big salon was empty.

I raced around the room a couple of times, bouncing the

ball in front of me as I went. My thoughts were carried far, far away. As with most children who grow up to themselves, a whole rich fantasy world had already formed inside me, the existence of which the grownups didn't begin to suspect.

I loved poetry with a passion. Its very form, its very rhythm delighted me. I greedily devoured every excerpt from Russian poets that caught my eye and, I have to confess, the more high-flown the poetry the better I liked it. For a long time the ballads of Zhukovsky were the only models of Russian verse I knew. No one in our house was particularly interested in this aspect of literature, and even though we had a good-sized library, it consisted mostly of foreign books. There was no Pushkin, no Lermontov, no Nekrasov. I could hardly wait for the day I first got my hands on a copy of Filonov's anthology,[1] purchased at the insistence of our tutor Malevich.

This book was a real revelation. For a few days afterward I went about like a lunatic, repeating to myself in an undertone stanzas from Lermontov's *The Novice* or Pushkin's *Captive of the Caucasus,* until Miss Smith threatened to take the precious book away from me.

The very beat of poetry enchanted me so much that I began composing verses myself at the age of five.[2] But my governess did not approve of this occupation. In her mind was formulated a perfectly crystallized concept of that normal, healthy child out of whom would later emerge the very model of an English miss. And poetry-writing did not accord with this concept at all.

For this reason she mercilessly hunted down all my poetic efforts. If, to my misfortune, she spied a scrap of paper with my doggerel scrawled on it, she would instantly pin it to my shoulder. And then, in the presence of my brother and sister, she would declaim the hapless piece out loud—twisting and distorting it cruelly, of course.

And yet none of this persecution had the least affect on me. By the time I was twelve, I was unshakably convinced that I was going to become a poet. Because of my fear of the governess I couldn't bring myself to write my verses down, but composed them inside my head like the bards of old, confiding them to my ball. Bouncing it in front of me, I would run through the salon while loudly reciting the two poetic works I was proudest of, "The Bedouin's Salutation to his Horse" and "The Sensations of a Pearldiver." Inside my head I had conceived a long epic, "The Streamlet," a kind of cross between *Undine* and *The Novice.*[3] But so far only the first ten stanzas had been finished, and the plan called for a hundred and twenty.

But the Muse, as we know, is capricious. Poetic inspiration did not always descend on me at the precise hour I was told to go and play ball. If the Muse failed to appear on summons my situation became dangerous, for I was surrounded by temptation on all sides. Next door to the salon was the library, and in that room the seductive volumes of foreign novels and issues of Russian periodicals were piled on all the tables and settees.

I was strictly forbidden to touch these, for my governess was highly selective on the subject of the reading considered permissible for me. I had a few children's books, and I already knew all of them almost by heart. Miss Smith never allowed me to read a book, not even a children's book, without making a preliminary check of it herself. And since she was a rather slow reader and never had any time for it, I was in a chronic state of book hunger.

And here, suddenly, at my very fingertips—such a treasure! How could anyone not be tempted?

I struggled with myself for a few minutes. I walked up to some book and at first I only glanced into it. I leafed through a few pages, read a few sentences and then started running along with my ball again, as if nothing had happened. But little by little the reading lured me back. Seeing that the first attempts had gone without mishap, I forgot all about the danger and started devouring page after page. It didn't even matter what it was—a novel of which this particular volume might not even be the first part—I still read it with just as much interest, starting from the middle and reconstructing the beginning in my own imagination.

Every now and then I would take the precaution of giving the ball a few bounces, so that the governess—just in case she had returned from her walk and had come to check on what I am doing—would hear me playing ball as I had been ordered to do.

Usually, this bit of cunning would work. I would hear her steps ascending the stairs in time and manage to tuck the book away to the side, so that she would be certain I had been amusing myself with the ball all this while, as was proper for a good, well-behaved child. But two or three times in my childhood it happened that I became so engrossed in reading that I never noticed anything until the moment when the governess rose up in front of me as if out of the earth, and caught me red-handed.

In situations of that sort, as with every particularly flagrant misdeed of mine, she would resort to her most extreme measure. She would send me to Papa with the order to tell him myself exactly what I had done. This punishment I feared more than any other.

In actuality our father was not at all strict with us, but I saw very little of him, only at dinner time. He never permitted himself the least familiarity with us except on those occasions when one of us was ill. Then he would change utterly. His fear of losing us made a different person of him. A rare tenderness and gentleness would come into his voice, his way of speaking to us. No one knew better than he how to show affection, how to joke with us. We truly adored him at such times and cherished these memories for a long time afterward.

But during "normal" times, when all of us were well, he kept to the rule that "the man of the house must be stern," and therefore he was very sparing with his caresses. He liked to be alone, and he had his own world which none of those at home were allowed to enter. In the mornings he went out walking on his managerial errands alone or with the estate steward, but all the rest of the day he spent inside his study.

This study, located quite far away from the other rooms, became a kind of holy of holies in the household. Even our mother never entered it without knocking first. As for the children, we would not dream of showing our faces there without an invitation to do so.

And so, when the governess said to me, "March yourself straight to your father and go boast to him how you've been behaving yourself!" I felt truly desperate. I wept, I resisted, but she was inexorable. Taking me by the hand she led, or rather dragged me through a long series of rooms to the door of Father's study, and there she left me to my fate and went away.

Now there was no more use crying. On top of that, the study was next to the entrance hall, and I had already caught a glimpse of an idle, curious lackey watching me with insulting interest.

"So you went and did it again, young lady!" the voice of Papa's valet Ilya, half-regretful and half-mocking, sounded behind my back.

I didn't vouchsafe him any answer and tried to put on an air of unconcern, as if I had come to visit Father of my own free will. To return to the classroom without obeying the governess's command was something I could not bring myself to do. That would mean redoubling my guilt by outright insubordination. And yet to go on standing there by the door, a target for the sneers of flunkeys, was unbearable. There was nothing left to do but knock on the door and bravely go inside to meet my fate.

I knocked, but very softly. A few seconds more went by, which seemed to me like an eternity.

"Knock louder, young lady! Your Papenka can't hear you!" again remarked the insufferable Ilya, who was evidently much taken with the whole episode. No way out of it: I knocked once more.

"Who's there? Come in!" sounded my father's voice at last from inside his study.

I entered, but stopped at the threshold in the half-darkness. My father was sitting at his writing desk with his back to the door and didn't see me.

"Well, who is it? What do you want?" he called out impatiently.

"It's me, Papa! Margarita Frantzevna sent me here!" I sobbed out in response.

Now, at last, my father guessed what it was all about.

"Aha! You've been bad again, no doubt!" he said, trying to give the sternest possible tone to his voice. "Well then, tell me the whole story! What have you done now?"

And so, sobbing and stammering, I began my denunciation of myself. Father heard me out absentmindedly. His concepts of child-rearing were very elementary, and the whole field of pedagogy seemed to him a female and not a male concern. It went without saying that he did not have the slightest intimation of the complex inner world which had already formed inside the mind of the young girl who now stood before him awaiting her sentence. Preoccupied with his "masculine" affairs, he hadn't even noticed how, little by little, I had outgrown the chubby little child I was five years ago. He was probably in a quandary about what it was that he was supposed to say to me and how he was supposed to behave in the given instance.

My misdeed seemed trivial to him, but he was a firm believer in the necessity of strictness in bringing up children. Inwardly he was annoyed with a governess who was not capable of handling so simple a matter by herself but had to send me to him. Now that his intervention had been sought, however, he was obliged to show his power. Therefore, so as not to undermine authority, he tried to put a severe and disapproving expression on his face.

"What a naughty child you are, you bad girl! I am very displeased with you," he said. Then he stopped because he didn't know what to say next. And since out of his entire fund of pedagogical wisdom the only thing his memory had retained was the notion that naughty children are made to stand in the corner, he decided, at last, "Go stand over there in the corner!"

What it felt like for me, a great big twelve-year-old girl who, only a few minutes before had been emotionally reliving the most complex psychological dramas together with the heroine of the novel she was reading clandestively—what it was like for me to be put to stand in the corner like an unreasonable little child, can be imagined.

Father went on working at his desk. Deep silence reigned in the room. I stood there without stirring, but God! The thoughts and the feelings that went on churning during those few minutes! I was so clearly aware of the ludicrousness and absurdity of the whole situation. Some sense of inner shame in front of my father forced me to obey him in silence and kept me from bursting into sobs or making a scene. But a feeling of bitter hurt and helpless rage rose up in my throat and choked me.

"This is all a pack of nonsense!" I tried inwardly to console myself. "A lot I care if I stand in the corner!" But just the same, it was painful to know that my father was able and willing to humiliate me. And this was the father I was so proud of, that I put higher than all others.

And still it wouldn't have been so awful if at least we were left alone. But now somebody knocked on the door, and the unbearable Ilya came into the room on some pretext or other. I knew perfectly well that this was a contrived reason, that he had come out of simple curiosity—he wanted to watch the young lady being punished! But he gave no sign at all, merely went about his business without hurrying, as if he hadn't noticed a thing. Only as he was leaving did he cast a sneering look my way. How I loathed him then!

I stood there so quietly that my father forgot all about me and therefore left me standing for quite a long time. I myself, of course would not ask his pardon for anything; my pride would not allow it. At long last he remembered that I was there and released me with the words, "Well, go on then, and look out you don't go playing any more tricks!"

It never occurred to him to imagine the emotional torment his little daughter had been suffering during that half hour. If he could look inside my heart he would probably have frightened himself. Within a few minutes, of course, he put this unpleasant childish episode completely out of his mind. But I left his study with a feeling of such unchildlike anguish, such undeserved insult, as has happened to me, perhaps, no more than two or three times since then, during the most painful moments of my life.

I returned to the classroom subdued and humbled. The governess was well pleased with the results of her pedagogical device, inasmuch as I was so quiet and so self-effacing for many days afterward that she could not praise my behavior enough. But she would have been less pleased if she had only known what an imprint was left on my spirit by this act of subjugation.

In general, the conviction that I was not loved in my family wound through my memories like a black thread. On top of the remarks of servants which I overheard accidentally, this lamentable belief was greatly reinforced by the isolated life I led with Miss Smith.

Nor was her own lot a happy one. Homely, lonely, no longer young, separated from English society but never quite assimilated into Russian life, she concentrated on me the whole force of her attachment, the whole craving for emotional possession that her abrupt, rigid, vigorous nature was capable of. I was truly the center and the focus of all her thoughts and concerns. I gave her life a sense of purpose, but her love for me was a heavy love, possessive, exacting and utterly without tenderness.

My mother and the governess were two such antithetical natures that it was not possible for any affinity to exist between them. In both personality and appearance, my mother belonged to that category of people who never age. She was a great deal younger than my father, and he went on treating her like a child almost up to old age. He addressed her as "Liza" or "Lizok," while she always gave him the dignity of the formal address, "Vasily Vasilievich." Sometimes he would even reprimand her in the presence of the children.

"You're talking rubbish again, Lizochka!" we often heard him say.

And Mama was not the least offended by such a remark.[4] But if she went on insisting on getting her own way, it was only in the manner of a spoiled child who has been allowed the right to want even what is unreasonable.

Mama positively feared Miss Smith, for the English-woman's caustic manner often hurt her feelings. Miss Smith deemed herself the exclusive proprietess of the children's rooms and received Mama as if she were a visitor. Therefore, Mama did not look in on us very often and didn't interfere in my upbringing.

In my own heart I was entranced by my mother, who seemed to me lovelier and more beautiful than any of the ladies of our circle. But at the same time the feeling of hurt never left me. Why was it that she cared less for me than for the other children?

I was sitting in the classroom one evening. My lessons for tomorrow were already finished and yet the governess, on this pretext or that, would not release me to go upstairs. And meanwhile from the floor above the sound of music carried to us from the drawing room, located directly over the classroom. It was Mama's habit to play the piano in the evenings. She could play for hours on end by heart, compose, improvise, go from one theme to the next. She had a high degree of musical taste and an amazingly pearly touch, and I dearly loved listening to her.

Under the spell of the music and my fatigue from the lessons just finished, a surge of tender feeling came over me, a need to press myself up to someone, cuddle close to someone. Only a few minutes were left before evening tea, and at last the governess let me go. I raced upstairs and came upon the following scene.

Mama had already stopped playing and was sitting on the couch. Anyuta and Fedya were snuggled up against her, one on each side. They were laughing, chattering away so animatedly that they didn't even notice me come into the room. I hung around them for a few minutes, saying nothing, hoping they would take notice of me. But they went right on talking about their own affairs.

That was enough to cool down all my enthusiasm. They didn't need me at all. A bitter, jealous feeling went through me, and instead of throwing myself at Mama and kissing her darling white hands as I had been fantasying while I was sitting downstairs in the classroom, I took myself off into a corner away from them, and I sulked there until we were called in to tea. Soon after that, I was sent away to bed.

CHAPTER SIX

My Uncle Pyotr Vasilievich Krukovsky

This conviction that my family loved me less than the other children disturbed me deeply, especially because the craving for an intense and exclusive attachment developed in me very early. The result was that the moment any of our relatives or friends showed a little more preference for me than for my brother or my sister, I immediately began to feel for that person an emotion verging on adoration.

I remember two particularly strong attachments in my childhood: to my two uncles. One of them was my father's elder brother, Pyotr Vasilievich Korvin-Krukovsky. He was a highly picturesque old man, tall, with a massive head and an aureole of very white and very thick curls. His face, with its correct, clearcut profile, shaggy gray eyebrows and deep vertical wrinkle cutting across his whole high forehead almost from bottom to top, might have appeared stern, almost harsh, if it had not been lit by such kindly, innocent eyes as are found only on Newfoundland dogs and small children.

This uncle was literally a man "not of this world."

Although he was the oldest member of our family and should have been its head, the truth was that he was ordered about by anyone who felt like it, and the whole family treated him like an elderly child. He had long held the reputation of an eccentric and a dreamer. His wife had died some years before; he had handed over his entire and rather good-sized estate to his only son,[1] leaving for himself only a very small monthy pension. Left thus without any definite business affairs to attend to, he used to come to Palibino often and stayed with us for weeks on end. His arrival was always regarded as a holiday, and the atmosphere at home became somehow livelier and cosier when he was with us.

His favorite corner was the library. He was extremely unwilling to exert himself physically and would sit immobile for days at a time on a big leather couch with one leg tucked under him, squinting his left eye (which was weaker than the right) and completely absorbed in reading the *Revue des Deux Mondes,* his favorite periodical.

Reading to the point of intoxication, to the point of stupefaction, was his only weakness. He was deeply concerned with politics. He would greedily devour the newspapers which arrived at our house once a week, and then he would sit for a long time and speculate: what new trick is that *canaille* Napoleoshka thinking up now?*

During the last years of Uncle's life, Bismarck also gave him a great deal of brain-wracking thought. He was certain, however, that "Napoleoshka will gobble up Bismarck" and, since he did not live to see the year 1870, he died still holding to his conviction.

On the subject of politics Uncle showed a startling bloodthirstiness. He was quite prepared to wipe out an army a hundred thousand strong. Nor was he less ruthless

*"Napoleoshka," a Russian pejorative for Napoleon III

in his imaginary punishment of criminals. To him, a criminal was a creature of fantasy, since in real life he saw all people as decent.

Notwithstanding the protests of our governess, Uncle sentenced all the British officials in India to death by hanging. "Yes, Madam, all, all!" he would shout, striking his fist against the table in the white heat of his passion. The expression on his face then was so menacing and ferocious that anyone coming into the room and seeing him would have been truly alarmed.

Then, abruptly, he would quiet down and a look of embarrassment and contrition would come over him, for he would suddenly become aware that this thoughtless gesture had frightened our general favorite, Grisi the greyhound, who was just settling herself next to Uncle on the couch.

But the thing that got Uncle worked up the most was coming upon a description in some periodical of an important new discovery in the field of science. On the days when he was with us there were heated arguments and discussions at the dinner table, whereas without him dinner usually took place in somber silence, since none of us, because of our lack of common interests, knew what to talk to each other about.

"And Sister, have you read what Paul Bert has come up with now?" he would ask, turning to my mother. "Made a pair of artificial Siamese twins. Spliced the nerves of one rabbit into the nerves of another one. If you hit one of them, the other one feels the pain. What do you say to that, eh? Do you realize what that could lead to?"[2]

And Uncle would start to relate the contents of the article he had just finished reading, embellishing and fleshing it out unwittingly, almost unconsciously, and deducing from it such bold conclusions and consequences as probably never entered the actual scientist's head.

When his account was finished, a hot discussion would flare up. Mama and Anyuta usually took Uncle's side right away and were full of enthusiasm for the new discovery. The governess, with her characteristic contrariness, almost as invariably placed herself in the ranks of the opposition and fiercely argued the irrationality or even the sinfulness of the theories voiced by Uncle. The tutor Malevich sometimes put in a word when the discussion turned on some purely factual point of information, but sensibly declined to take any direct part in the dispute. As for Father, he assumed the role of skeptical, gibing critic who sides with neither faction but only points out and clearly underscores all the weak points of both camps.

These disputes sometimes took on a distinctly bellicose character. By some fateful chance, it almost always happened that questions of a purely abstract nature would unexpectedly shift into the sphere of petty personal sarcasms. It was always Miss Snith and Anyuta who were the bitterest adversaries. A muted Seven-Years-War was going on between them, interupted only by periods of armed and wary truce.

If Uncle stunned us with the boldness of his generalizations, then the governess on her part was distinguished in the sphere of applied knowledge. In the most abstract and apparently remote scientific hypotheses, she would suddenly discern grounds for criticizing Anyuta's behavior, such unexpected and original ones that the others could do no more than shrug their shoulders helplessly.

Anyuta gave as good as she got, and answered back so angrily and rudely that the governess would jump up from the table and announce that she would not remain in our house after such an insult. All those present felt awkward and uncomfortable. Mama, who loathed quarrels and scenes, took upon herself the role of mediator, and after protracted negotiations peace was concluded.

I still remember the storm raised in our household by two articles in the *Revue des Deux Mondes.* One of them was a review of Herman Helmholtz's publication on the unity of physical forces,[3] the other an account of Claude Bernard's experiments in the resection of parts of the brain in a pigeon. Both Helmholtz and Bernard would doubtless have been astonished to learn what an apple of discord they had cast into a peaceful Russian family living somewhere in the wilds of the province of Vitebsk.

But it was not only politics and reports of the latest scientific discoveries which had the power to agitate Uncle Pyotr. With equal zeal he read novels and travel accounts and historical articles. In the absence of anything better, he was prepared to read even children's books. I have never found in anyone, except perhaps in certain adolescents, such a passion for reading as he had.

It would seem that nothing could be more innocent than such a passion, nor simpler for a wealthy landowner to satisfy. But, as it happened, Uncle had almost no books of his own, and it was only during the last years of his life, and that thanks to our Palibino library, that he was able to indulge himself in the only pleasure he prized.

Owing to an unusual character weakness which was completely at variance with his stern and majestic appearance, Uncle found himself under someone's domination all his life—a domination so cruel and autocratic, moreover, that it was out of the question for him to indulge any whim or personal taste. As a result of this same character weakness, he had been adjudged from childhood as being unfit for military service, the only occupation considered respectable at that time for a hereditary nobleman. And since he was a docile child by nature, not given to mischief, his tenderhearted parents decided to keep him at home, giving him only enough education to exempt him from military service in the ranks.[4]

Everything Uncle knew, he had either figured out for himself or read about later in books. And yet his store of knowledge was truly remarkable although, as with all self-taught people, it was scattered and uneven. On one subject he displayed great erudition, in another, scarcely any information at all.

When he reached adulthood he continued to live at home in the country, evincing not the slightest self-esteem, content to hold the humblest position in the family. His younger and far more brilliant brothers behaved arrogantly towards him, with a kind of good-natured condescension, as if to a harmless eccentric.

Then suddenly fortune descended upon him like a bolt from the blue. The leading beauty and richest match in the whole province, Nadezhda Andreyevna N., turned her attentions upon Uncle. Whether she was infatuated with his good looks or had simply calculated that he would be just the kind of husband she needed, that it would be nice to have this tall, docile, devoted creature always at her pretty little feet—the Lord only knows. Be that as it may, she gave him clearly to understand that if he proposed to her, she would be glad to marry him.

Uncle Pyotr himself would never have dared even to dream of anything of the sort. But his numerous aunties and other female relations were quick to clarify the good fortune that had fallen to his lot, and before he realized what was happening to him he was betrothed to the beautiful, self-willed and spoiled Nadezhda Andreyevna. But no happiness came out of this union.

We children were quite certain that our Uncle Pyotr existed on earth primarily for our pleasure. Without the least hesitation we would babble to him every piece of nonsense that came into our heads. And yet all of us, as if by some instinct, sensed that there was one topic that was not to be touched upon: one must never ask Uncle any questions about his late wife.

Among ourselves the most sinister stories circulated about Aunt Nadezhda Andreyevna. The grownups—that is, Father, Mother and the governess—never so much as mentioned her name in our presence. But once in a while

Aunt Anna Vasilievna, my father's younger unmarried sister, found herself in a loquacious mood, and then she would tell us different horror stories about "my late sister Nadezhda Andreyevna."

"That was a real viper, that one! God preserve us! She ate us up alive, me and sister Marfinka! And didn't brother Pyotr have to take it from her, too! There would be times when she would get angry at one of the servants, and she'd run straight to brother Pyotr's study and demand that he chastise the guilty party with his own hands. He, because of his kindness, didn't want to do it, he'd try to reason with her—hopeless! The more he reasoned, the more furious she'd get. Then she'd turn on him and start abusing him with all kinds of vile language. She'd call him a lazy sloth, no resemblance to a real man. . . . It made you feel ashamed just listening to it.

"Finally she'd see that words weren't getting her anywhere with him. So she'd snatch up a handful of his papers, his books, whatever came to hand on his writing desk and she'd throw it all into the stove. 'I won't have this trash in my house!' she'd shout.

"Sometimes it would even happen that she'd take the slipper off her foot and then take and slap his face with it! Yes—really! And go on slapping. And he didn't do a thing, my poor dear, just tried to hold her hands back, and so carefully, so as not to hurt her, and tried to reason with her, ever so gently: 'What are you doing, Nadenka, calm down! Aren't you ashamed? And in front of other people, too!' But she had no shame, that one."

"But how could Uncle bear to be treated like that?" we exclaimed, outraged. "Why didn't he leave her?"

"Oh, my darlings, you don't think a person can shake off a lawful wedded wife, now, like a glove?" Aunt Anna replied. "And anyway, I had better tell you that no matter how she ordered him about, he was madly in love with her all the same."

"Did he really love her? A mean woman like that!"

"Love her! And how he loved her, children—he couldn't live without her! And when they did her in, he took it so hard—he almost laid hands on himself."

"What in the world are you saying, Auntie? Did her in? What's that?" we asked curiously.

But Auntie, having realized that she has told too much, suddenly broke off her story and started energetically knitting a stocking so as to show us there would be no continuation. Our curiosity was inflamed by now, however, and we didn't let up on her.

"Auntie, darling, tell us, tell us!" we kept at her.

Auntie herself, evidently, had worked herself into a talking mood and could not stop.

"Well . . . this is how it was . . ." she blurted out. "Her own serf girls strangled her."

"God! How horrible! How did it happen? Auntie, dearest Auntie, tell us!" we cried.

"Just so—very simple!" Aunt Anna related. "One night, she was all alone in the house. She'd sent brother Pyotr and the children off somewhere. That night her favorite maid Malanya undressed her and took off her shoes and tucked her into her bed, all right and proper. And then, suddenly—she claps her hands! At this sign, the rest of the maids come trooping in to the bedroom from all the other rooms nearby, and the coachman Fyodor too, and the gardener Evstigney. Sister Nadezhda Andreyevna, the moment she gets a look at their faces, she realizes right away that something's wrong. But she doesn't get scared, she doesn't lose her head. She yells at them, 'Where do you think you're going, damn you! You're out of your minds! Get out of here this minute!'

"They were so used to obeying that they were frightened and they were already backing away toward the door when Malanya—she was bolder than the rest of them—began talking them back into it. 'What's the matter with you, you low cowards, you? You're not afraid for your own hides, is that it? Why, tomorrow she'll pack the whole lot of you off to Siberia!'

"Well, finally, they understood what was up. And the whole mob of them, they tramped over to her bed, they grabbed our late sister by the hands and feet, they stuffed a feather bed over her head and started strangling her. And she kept begging them, and promising them money and all kinds of things. No—it was too late to do anything with them.

"And Malanya, her favorite, kept urging them on: 'It's a towel you need, a wet towel, throw it on her head so there won't be any black and blue marks on her face.'

"And then they themselves, the vile serfs, went afterwards and confessed. At the trial, under the whip, they told all the details, how it all happened. Well, then! They didn't get a pat on the head for that, for that good deed of theirs. There are plenty of them, I expect, still rotting in Siberia to this very day!"

Aunt Anna fell silent, and we children, aghast, were also silent.

"Take care now, don't go telling your Papenka or your Mamenka how I let out the whole story without thinking!" Auntie admonished us. But we understood ourselves that such things were not talked about, not with Papa nor with Mama nor with the governess. It would only cause an uproar.

All the same, that night when the time came for bed, the story haunted me and wouldn't let me fall asleep. Once, on a visit to Uncle Pyotr's estate, I had seen a portrait of Aunt Nadezhda Andreyevna painted in oils, a full-length portrait done in the usual conventional manner in which portraits were painted at that time. And now I imagined her before me, a living person. Short, elegant as a china doll, in a low-cut crimson velvet dress, with a garnet necklace lying across her ample white breast, with bright red color on her round cheeks, with an arrogant expression in her big dark eyes and a stereotyped smile on her rosy little mouth. And I tried to picture those big eyes growing even bigger, the dread that was in them when suddenly she saw her docile slaves, come to murder her!

Then I started imagining that I myself was in her place. While Dunyasha was undressing me, the thought suddenly occurred to me—what if that kindly round face of hers should suddenly change and grow evil? What if she should suddenly clap her hands, and Ilya and Stepan and Sasha would come into the room and say, "We came to murder you, miss!"

All at once I was really terrified by this absurd notion, so that I didn't hold Dunyasha back as usual but, on the contrary, was almost glad when at last, after finishing my preparations for the night, she went away and took the candle with her. But I still couldn't fall asleep. For a long time I lay there in the dark with my eyes open, waiting impatiently for the governess, who had stayed upstairs to play cards with the grownups, to come to my room.

After that, every time I was alone with Uncle Pyotr this story would return to my mind of its own accord; and it seemed to me odd and inexplicable that this man, who had suffered so deeply in his life, could now so serenely, as though nothing had ever happened, be playing chess with me, or building me little paper ships, or getting all worked up about some project for the restoration of the ancient bed of the Syr-Darya River or about some other newspaper article he had just read. It is always so hard for children to imagine that anyone close to them, whom they are accustomed to seeing every day in ordinary life, can have experienced an extraordinary, tragic event in his own lifetime.

There were times when I was simply seized with a morbid desire to ask Uncle how it all happened. I would stare at him for a long time and then, without taking my eyes off him, I would picture this tall, strong, clever man trembling before his tiny beauty of a wife, crying and kissing her hands, while she would tear up his papers and his books or take off her slipper and start slapping him across the face.

Once, only once in all my childhood, I couldn't hold back any longer, and I put my finger on Uncle's sore spot. It was in the evening. We were alone in the library. Uncle, as usual, was sitting on the couch and reading with one leg tucked under him. I was running around the room bouncing my ball. Finally, tired out, I sat down on the couch beside him. Staring at him, I surrendered to my old fantasies about him.

Suddenly Uncle put his book down, stroked my hair tenderly and asked, "What are you thinking about, child?"

"Uncle, were you very unhappy with your wife?" burst out of me, almost against my will.

I shall never forget the impact of this unexpected question on poor Uncle. His quiet, stern face broke up into tiny wrinkles as though with physical pain. He even stretched out his hands in front of him as if warding off a blow. And I felt so sorry for him, so pained and so ashamed—as if I too had taken the slipper off my foot and slapped his face.

"Uncle darling—forgive me! I wasn't thinking!" I said, pressing close to him and hiding my face, red with shame, against his chest. And so my kindly uncle then had to console me for my indiscretion.

From that time on I never again returned to this forbidden subject. But I could ask him direct questions about anything else. I was considered his favorite, and we spent hours together discussing all kinds of things. When he was preoccupied with some idea, it was the only thing he could talk and think about. Forgetting completely that he was talking to a child, he often developed the most abstract theories before me. And that was exactly what I enjoyed most, that he spoke to me as though to an adult, and I concentrated all my efforts to understand him or at least to give the appearance of understanding him.

Although he had never studied mathematics he had the most profound respect for that branch of learning. From different books he had accumulated some smattering of mathematical knowledge and loved to philosophize about it, and he often reflected aloud in my presence. It was from him, for example, that I heard for the first time about squaring the circle, about the asymptote, toward which a curve approaches constantly without ever reaching it, about many other matters of a similar nature. The meaning of these concepts I naturally could not yet grasp, but they acted on my imagination, instilling in me a reverence for mathematics as an exalted and mysterious science which opens up to its initiates a new world of wonders, inaccessible to ordinary mortals.

As I speak of these, my first contacts with mathematics, I cannot help mentioning a curious circumstance which also helped to arouse my interest in the field.

When we moved permanently to the country, the whole house had to be redecorated and all the rooms had to be freshly wallpapered. But since there were many rooms, there wasn't enough wallpaper for one of the nursery rooms. Because ordering wallpaper involved sending to Petersburg it was a very complicated business, and it really wasn't worth while to go through all of that for just one room. It was all waiting for a propitious occasion, and in expectation of this the maltreated room just stood there for many years with one of its walls covered with ordinary paper. But by happy chance, the paper for this preparatory covering consisted of the lithographed lectures of Professor Ostrogradsky[5] on differential and integral calculus, which my father had acquired as a young man.

These sheets, all speckled over with strange, unintelligible formulas, soon attracted my attention. I remember as a child standing for hours on end in front of this mysterious wall, trying to figure out at least some isolated sentences[6] and to find the sequence in which the sheets should follow one another. From this protracted daily contemplation, the outer appearance of many of these formulas imprinted themselves in my memory; indeed, their very text left a deep trace in my brain, although they were incomprehensible to me while I was reading them.

Many years later, when I was already fifteen[7] I took my first lesson in differential calculus from the eminent Petersburg professor Alexander Nikolayevich Strannolyubsky.[8] He was amazed at the speed with which I grasped and assimilated the concepts of limit and of derivatives, "exactly as if you knew them in advance." I recall that he expressed himself in just those words. And, as a matter of fact, at the moment when he was explaining these concepts I suddenly had a vivid memory of all this, written on the memorable sheets of Ostrogradsky; and the concept of limit appeared to me as an old friend.

CHAPTER SEVEN

Uncle Fyodor Fyodorovich Shubert

My attachment to my other uncle, my mother's brother Fyodor Fyodorovich Shubert,[1] was of quite a different order. This uncle, the only son of his late father, was a good deal younger than my mother. He lived permanently in Petersburg and, as the only male representative of the Shubert family, he enjoyed the boundless adoration of his sisters as well as of his numerous maiden aunts and cousins.

A visit from Uncle to us in the country was looked on as a real event. I was about nine when he came to us for the first time. His coming had been discussed for many weeks in advance. The best room in the house was given to him, and Mama herself saw to it that the most comfortable furniture was put into it. We sent a carriage to meet him at the provincial capital, one hundred and fifty versts distant from our estate, and fitted it out with a fur greatcoat, a fur rug and a laprobe, so that Uncle might not catch cold, since it was already late in autumn.

On the evening before the day of Uncle's expected arrival we suddenly looked out and saw driving up to the front entrance an ordinary cart hitched up to a troika of old post-station nags. A young man jumped out of it dressed in a light city topcoat, with a leather traveling bag flung over his shoulder.

"My Lord! There's brother Fedya!" exclaimed Mama, looking out of the window.

"Uncle's here! Uncle's here!" rang out through the house, and we all ran into the front hall to greet our guest.

"Fedya, you poor thing!" said Mama in a commiserating tone, throwing her arms around her brother. "How come you used the post horses? Didn't you meet the carriage we sent for you? You must be all shaken up!"

It seemed that Uncle had left Petersburg a day earlier than expected.

"For goodness sake, Liza!" he said, laughing and wiping the drops of frost from his mustache before kissing his sister. "I had no idea you'd make such a fuss over my coming! What was the point of sending a carriage for me? Am I an old lady, that I can't drive a hundred and fifty versts in a cart?"

Uncle spoke in a very pleasant, deep tenor voice, rolling his r's in the Petersburg style. He looked quite young. His short-cropped chestnut hair lay on his head like thick, velvety beaver fur, his pink cheeks were shiny with frost, his dark brown eyes had an enthusiastic, alive expression, and a row of strong white teeth showed from behind his full red lips, bordered by an elegant mustache.

I looked him over rapturously. "What a fine man, this uncle of mine! He's gorgeous!" I thought.

"And who's that?" Uncle asked, pointing to me. "Anyuta?"

"Come now, Fedya—Anyuta's a young lady by now," Mama corrected him in an offended tone. "That's only Sonya!"

"My lord, your daughters have really grown, haven't they? Look out, Liza, before you realize what's happening they'll put you down as an old lady," Uncle said with a laugh, and he kissed me. I couldn't help feeling embarrassed by his kiss and turned red all over.

At dinner, of course, Uncle occupied the place of honor next to Mama. He ate with a hearty appetite, which didn't hinder him in the least from talking without a stop. He related various news items and tidbits of gossip from Petersburg, made us all laugh and went into fits of ringing, merry laughter himself. All of us listened to him very attentively. Even Papa treated him with great respect, without a trace of that haughty, half-patronizing, half-mocking manner he so often assumed with the young relatives who came to visit us, and which they so disliked.

The more I looked at my uncle the better I liked him. He had already managed to wash and change, and one would never have guessed from his fresh, healthy look that he had just arrived after a long journey. His jacket, of a thick, tweedy English cloth, fit him extremely well, not at all the way other people's jackets fit. But what I liked most of all were his hands—large, white, well groomed, with gleaming fingernails like large pink almonds. I didn't take my eyes off him all through dinner and even forgot to eat, I was so busy looking him over.

Preserved gooseberries were served for dessert. Uncle put a large portion on his plate; the big green berries floating in their thick syrup looked so delicious. He looked at the preserves, then at me, and then at the fruit again, and suddenly burst out with such a gay, infectious laugh that everybody else started laughing too without knowing why.

"You know, Liza," Uncle said at last, suppressing his laughter with an effort, "all during dinner I've been sitting and wondering what Sonya's eyes remind me of. And now I know: they're like the gooseberries in these preserves, just as big, just as green and sweet . . ."

All found this an uncommonly apt comparison and greeted it with a new outburst of laughter. I blushed to my ears and was prepared to feel insulted, but then Uncle added, laughing, "but very beautiful and very green." And this mollified me a little.[2]

After dinner Uncle sat down on a small corner settee in the drawing room and put me on his lap.

"Come on, Mademoiselle Niece," he said. "Let's get acquainted." And he began asking me all kinds of questions: what I was studying, what I was reading. Children usually know their own strong and weak points much better than grownups realize. Thus, for example, I was fully aware that I was doing well in my studies and that everyone considered me very *avancée* for my age.

Therefore I was quite pleased that it had occurred to Uncle to ask me about this, and I answered all his questions freely and willingly. And I saw that Uncle was pleased with me.

"What a clever one!" he kept repeating every minute. "Imagine her knowing all that already!"

And then I begged in my turn, "Uncle, now you tell me something!"

"Well, all right, if you like. But a person can't tell fairy tales to such an intelligent young lady as you. With you one can speak only on serious subjects."

And he began telling me about infusoria, about algae, about the formation of coral reefs. Uncle himself had graduated from the university not so very long ago, so that all this information was fresh in his memory. He spoke very well, and he was pleased to have me listen so attentively, with my eyes open wide and riveted on him.

After that first day the same routine would be repeated every evening. Mama and Papa would both go off for a half hour's nap after dinner. Uncle had nothing to do. He would sit down on my favorite settee, take me on his knees and start talking to me about everything under the sun. He invited the other children to take part, too. But my sister, who had only recently risen from the school bench herself, was afraid of lowering her dignity as a grown-up young lady if she started listening to such edifying things, "interesting only to little children." And my brother once stood there and listened for a little while, found it boring and ran away to play horsie.

But as for me, our "scholarly discussions," as Uncle jokingly called them, became precious beyond words. My favorite time in the whole day was the half hour after dinner when I was alone with him. I genuinely adored him. To speak frankly, I can't even vouch that this feeling was not mingled with a kind of childish lovesickness, an emotion for which small girls have much more capacity than adults realize.

I felt a strange embarrassment whenever I said his name, even if only to ask, "Is Uncle at home?" If anyone at the dinner table, noticing that I never took my eyes off him, made a remark like "So, Sonya, one can see that you love your uncle a lot?" I would blush fiery red and not answer.

I saw almost nothing of him during the day, for my life was quite apart from the life of the grownups. But during lessons and recreation I thought of one thing only: "Will evening come soon? Will I be with Uncle soon?"

One day some neighboring landowners who had a daughter, Olya, came for a visit while Uncle Fyodor was staying with us. This Olya was the only other child my own age whom I ever had the opportunity to meet. She wasn't brought to our house very often, but when she was, it would be for all day. Sometimes she even stayed overnight.

She was a very lively, merry little girl. Our personalities and tastes were quite different, so that no real friendship existed between us. Still, I would usually be very happy to see her, especially because I would be excused from my lessions and given an all-day holiday in Olya's honor.

But this time the first thought that flashed into my head when I saw Olya was: what will happen after dinner? The chief charm of my chats with Uncle consisted precisely in being alone with him, in having him all to myself. And I already sensed that silly Olya's presence would spoil everything.

Therefore I greeted my little girlfriend with a good deal less pleasure than usual. All morning I kept hoping secretly, "Maybe they'll take her home earlier today?"

But no: as it turned out, Olya was to remain till late in the evening. What to do? With some misgivings, I decided to be frank with my friend and ask her not to get in my way.

"Look here, Olya," I said furtively, "I'll play with you all day, and I'll do whatever you want me to do. But then please do me a favor and go away somewhere after dinner and leave me alone. My uncle and I always have a conversation after dinner, and we don't need you there at all!"

To this Olya agreed, and throughout that day I fulfilled my part of the bargain honestly. I played all the games she could think up, acted all the parts she wanted me to act, switched from being a ladyship to being a cook and back again from cook to ladyship the moment she ordered me to do so.

At last we were called in to dinner. I sat at the table on tenterhooks. I kept wondering whether Olya would keep her promise, and from time to time I would steal an uneasy glance at her, trying with meaningful looks to remind her of our bargain.

When dinner was over I went up to Papa and Mama as usual to kiss their hands and then hung around Uncle Fyodor, waiting to hear what he would say.

"Well, how about it, my girl, are we going to have our discussion today?" he asked, giving me an affectionate chuck on the chin.

I literally jumped for joy. Happily clutching his hand, I was all ready to settle down in our cherished corner. But then I suddenly noticed that treacherous Olya was following right behind us.

As it turned out, my prior arrangements had only managed to spoil everything. It is quite possible that if I had told her nothing she herself, having a healthy aversion to anything reminiscent of lessons and seeing that Uncle and I were about to have a talk about serious things, might have run off of her own accord. But when she saw how I treasured Uncle's stories and how anxious I was to be rid of her at all costs, she imagined that we would be talking about something really fascinating, and she too wanted to be part of it.

"But can I come along with you too?" she asked imploringly, raising her appealing blue eyes to Uncle.

"Of course you can, sweetheart," he replied, glancing fondly at her and oviously admiring her pretty, rosy little face.

I threw Olya an irate, furious look, which did not disconcert her at all. "But Olya doesn't know about these things. She won't understand any of it anyway," I resentfully tried to point out.

But this attempt to get rid of my persistent friend had no success either.

"Well, in that case we'll talk about simpler things today, so that it will be interesting for Olya too," Uncle said goodnaturedly. And, taking us both by the hand, he walked us toward the settee.

I went in sullen silence. A three-sided conversation, with Uncle talking for Olya's benefit, adjusting to her tastes and understanding, was not at all what I had in mind. I felt that something had been taken away from me that belonged to me by right, something inviolable and precious.

Uncle, who was apparently not even aware of my bad humor, said to me, "Well then, Sonya, climb up on my lap!"

But I was so insulted that this invitation didn't propitiate me at all.

"I don't want to!" I answered angrily, and went off into a corner to sulk.

Uncle looked at me with an expression of surprise and amusement. Whether he was aware of the jealousy stirring inside me and felt like teasing me a little, I don't know; but he turned suddenly to Olya and said, "All right then—if Sonya doesn't want to sit on my lap, you do it!"

Olya didn't need to be asked twice. Before I even realized what was happening she was already occupying my place on Uncle's knee. That was something I hadn't expected at all! It hadn't occurred to me that matters could take such a dreadful turn. It literally seemed that the earth was giving way under my feet.

I was too stunned to express any protest. All I could do was to stare silent and wide-eyed at my lucky friend. But she, just the least bit embarrassed but very pleased all the same, had settled herself on Uncle's knee as if everything were quite in order. Arranging her little mouth into a comical moue, she tried very hard to put on her round child's face an expression of seriousness and attention. She was blushing all over. Even her neck and her bare little arms were scarlet.

I looked at her, I looked . . . and then, suddenly—even now, I swear, I don't understand how it came about—something awful happened. It was as though someone else was pushing me. Without a conscious thought of what I was doing, I suddenly, to my own amazement, sank my teeth into her bare, chubby little arm, slightly above the elbow, and I bit her till the blood came.

My attack was so startling, so unforeseen, that for the first second all three of us were stunned and only stood wordlessly staring at each other. Then Olya suddenly let out a piercing shriek, and that shriek brought us back to reality.

Shame. Bitter, desperate shame overwhelmed me. I fled headlong from the room and heard my uncle's indignant voice saying as I left, "Nasty, mean brat!"

My time-honored haven during all the major troubles of my life was the room which had belonged to our former seamstress Marya Vasilievna and which had been given to

Nanny. And there I now went for refuge. Burying my head in the kindly old woman's lap, I sobbed long and hard. Nanny, seeing me in such a state, asked no questions but only stroked my hair and lavished pet names on me. "God bless you, my darling, calm down, my sweetie," she said, and in my overwrought state it felt so good to cry it all out in her lap!

Luckily, the governess was not at home that night. She had gone on a few days' visit to some neighbors. Therefore nobody noticed that I was missing, and I could cry to my heart's content in Nanny's room. After I had quieted down a little she fed me some tea and put me to bed, where I immediately fell into a deep, leaden sleep. But when I woke up the next morning and suddenly remembered the night before, I again felt so ashamed that I thought I could never again bring myself to look anyone in the face.

The whole episode turned out far better than I could have expected. Olya had been taken home the night before. Evidently she had been high-minded enough not to tell on me. It was evident by the look on people's faces that nobody knew a thing. No one scolded me for yesterday's incident, no one teased me. Even Uncle behaved as though nothing out of the ordinary had happened.

Strange to relate, however, my feelings toward my Uncle Fyodor changed completely from that day on. Our after-dinner conversations were never resumed again. Soon after that he returned to Petersburg, and although we often met later, and he always treated me very kindly, and I was very fond of him—still, the old adoration of him was something I never felt again.

CHAPTER EIGHT

My Sister

But incomparably stronger than all the other influences which affected my childhood was the influence of my sister Anyuta. The emotion I felt for her from my earliest years was a very complicated one. I admired her beyond bounds, yielded to her unquestioningly in all things and felt highly flattered whenever she allowed me to take part in anything with which she herself was concerned.

I would have gone through fire and water for my sister. But at the same time, in spite of my fierce attachment to her, a grain of envy nestled in the depths of my heart—that special envy we so often, almost unconsciously, harbor for people to whom we feel very close, whom we admire very deeply and would like to emulate in all things.

And yet it would be wrong to envy my sister for, if the truth be known, her destiny was far from happy.

My parents moved permanently to the country just at the time when Anyuta was beginning to grow into adolescence. The Polish uprising erupted shortly after our move.[1] Since our estate lay very close to the Russian-Lithuanian border, the reverberations of this uprising reached us as well. Most of the neighboring gentry, and preeminently the richest and most highly cultivated ones, were Poles. Many of them were more or less seriously compromised politically. Some of them had their estates confiscated; almost all were forced to pay heavy indemnities. Many abandoned their estates of their own accord and went to live abroad.

Somehow, there were almost no young people to be found in our parts in the years following the Polish uprising; they had all disappeared somewhere. Only children and old people—harmless, intimidated, afraid of their own shadows—were left. And there were also new arrivals of various kinds: officials, merchants and petty gentry.

It is understandable that under conditions such as these rural life was not particularly gay for a young girl.[2] Moreover, all of Anyuta's early upbringing had been of a kind that could not cultivate any taste in her for country living. She did not like going out for walks, nor gathering mushrooms, nor rowing. To make things worse, the organizer of all such pastimes was invariably our English governess. And the antagonism between her and Anyuta was so intense that the moment one of them made any suggestion the other would hostilely reject it.

It is true that Anyuta conceived a passion for horseback riding one summer, but that seems to have been mostly from imitation of the heroine of some novel in which she was then engrossed. Since no suitable companion was available she quickly lost interest in these solitary excursions on horseback, accompanied only by our bored coachman; and her horse, whom she had christened with the romantic name of "Frieda," was soon transferred to the more modest work of taking our estate steward to the fields and reverted once more to her original name of "Golubka."

The possibility of my sister's taking some part in running the household was out of the question. Such a suggestion would have appeared ludicrous both to her and to those about her. Her entire upbringing had been focused on making a brilliant, sophisticated young lady out of her.

Practically from the age of seven she had been accustomed to being the queen of all the children's balls, to which she was often taken while our parents were living in large cities. Papa was very proud of her childish successes, many legends of which circulated in our family.

"Our Anyuta, when she grows up, she'll be fit for a royal court—any prince would lose his mind over her," Papa used to say, as a joke, of course. But the trouble was that not only we, the younger children, but also Anyuta herself took these words in dead earnest.

My sister was very lovely in her early youth. Tall, slender, with a flawless complexion and a mass of fair hair, she was virtually a raving beauty; and besides that she had a great deal of her own special charm. She was perfectly well aware that she was capable of playing the leading role in any social group. And now, suddenly—the countryside, the wilds, tedium.

She would often go to Father and reproach him, with tears in her eyes, for making her live in the country. At first he only tried to laugh her out of it, but sometimes he would condescend to explanations and try to show her very logically that at this difficult time it was the duty of every landowner to live on his own estate. To abandon the estate now would mean the ruin of the whole family.

Anyuta did not know how to reply to these arguments. She only felt that knowing them didn't make her feel any better, that her youth would not be repeated a second time. After conversations of this sort she would go off to her own room and weep bitterly.

But once a year, in winter, Father would send my mother and sister to Petersburg for a month or six weeks, to stay at our aunts'. These excursions, which cost huge sums of money, were actually of no help. They merely inflamed in Anyuta the taste for pleasure while offering no real gratifications. The month in Petersburg would go by so swiftly that she had no time to take it all in.

The kind of person who might have directed her mind to serious subjects was not to be found in the social circles to which she was taken. Nor were eligible suitors available. An elegant wardrobe would be made for her, I remember. She would be taken to the theater three or four times, or to a ball at the Nobility Hall. One of the relatives would give a dinner party in her honor; she would be showered with compliments on her beauty.

And then, just as she was beginning to develop a real taste for all this, they would take her back to Palibino again. And again would begin for her the isolation, idleness, tedium, the pacing back and forth for hours on end in the vast rooms of our Palibino house, reliving in fantasy her recent pleasures and her feeble, sterile dreams of new social successes.

In an effort to find some way to fill the emptiness of her life, my sister was continually inventing artifical diversions. And, since the life of those at home was very bare of inner content, every one used to fall upon each new project of hers fervently as a pretext for discussions and excitement. Some criticized her, others were on her side, but she provided a pleasant interruption for all in the monotonous routine of their lives.

When Anyuta was only fifteen years old, she displayed her first act of independence by pouncing upon all the novels to be found in our home library, and consuming an incredible quantity of them. Fortunately, there were no "bad" novels in our house, although there was no lack of badly written and untalented ones. Our library's chief riches, however, consisted in a great number of old English novels, mostly historical ones in which the action took place in the middle ages during the period when knighthood was in flower.

For my sister, these novels were a true revelation. They carried her away into a world of wonders unfathomed up to then, and they gave a new direction to her imagination. The same pheonmenon that had occurred to Don Quixote many centuries before repeated itself with her: she came to believe in knights and imagined herself a medieval lady.

As bad luck would have it, our huge, massive country manor, with its tower and its Gothic windows, was constructed somewhat in the style of a medieval castle. During her knightly phase Anyuta could not write a single letter without putting on it the heading: Château Palibino. The top room of the tower, which had long remained in disuse, so that even the steep rungs of the staircase leading to it were mildewed and rickety, she ordered cleaned of dust and spiderwebs. She hung the walls with old rugs and weapons unearthed somewhere from the attic rubbish, and converted the tower into her permanent residence. I can still see her graceful, slender figure tightly draped in a white dress, and with two heavy blonde braids hanging below her waist. In this costume Anyuta would sit at a tambour frame and embroider the family crest of King Matthias Corvinus in beads on canvas, looking out of the window toward the highroad from time to time to see whether a knight might be coming.

"Sister Anne, Sister Anne! Don't you see someone coming?"

"I see nought but the dusty earth and the greening grass."

Instead of the knight, a district police officer would drive up, or some excise officials, or some Jews come to buy up Father's vodka and oxen. But there were no knights.

Finally she grew bored with wishing for a knight, and the courtly phase passed just as quickly as it had begun.

Just at the time when she was beginning, still unconsciously, to find courtly novels less than satisfying, a highly exalted novel came into her hands. It was Edward Bulwer-Lytton's *Harold, the Last of the Saxon Kings.*

After the Battle of Hastings, so the story went, Edith of the Swan's Neck discovered the body of King Harold, her betrothed, among the dead. Just before the battle he had broken an oath—a mortal sin—and died without the opportunity to repent. His soul was condemned to eternal torment.

After that day Edith too disappeared from her native land, and none of her kinfolk ever heard more of her. Many years passed, and the very memory of Edith began to be forgotten.

But on the opposite coast of England, amid wild cliffs and forests, stood a convent renowed for the austerity of its regime. There, for many years, lived a certain nun who had taken a vow of eternal silence and aroused the admiration of all the other nuns through her feats of piety. Neither by day nor by night did she know rest; in the early hours of the morning and the dead of night her genuflecting figure could be seen before the crucifix in the convent chapel. Wherever a duty existed to be performed, help to be proffered, another's suffering to be consoled—everywhere she was first on the scene. Not a soul in the district passed on without the tall figure of the pale nun bending over his deathbed, without his forehead, already covered with the cold sweat which heralds death, being touched by her bloodless lips, sealed in the dread vow of eternal silence.

And yet no one knew who she was and whence she had come. Twenty years before or more, a woman wrapped in a black veil had appeared and, after a prolonged and secret discussion with the abbess, had remained permanently. That abbess had died long before. The pale nun went wandering like a shadow, but none of those now alive in the convent had ever heard the sound of her voice.

The young nun and the poor folk in all the countryside around revered her like a saint. Mothers brought their sick children to her to be touched by her hand, in the hope of being healed by her touch alone. But there were also those who spread the rumor that, if she must redeem her past through such self-flagellation, she must, indeed, have been a great sinner in her youth.

At last, after long years of sacrificial labor, the hour of the nun's death arrived. All the nuns, young and old, crowded about her death bed. The Mother Abbess herself, long deprived of the use of her legs, commanded that she be carried into the nun's cell.

And now the priest entered. By the power vested in him by our Lord Jesus Christ, he released the dying woman from her self-imposed vow of silence and charged her to confess before she died who she was, what was her sin, what crime weighed heavy upon her conscience. With an effort, the dying woman raised herself on her pallet. Her bloodless lips, grown unused to human speech, seemed to have turned to stone during their long silence. They moved convulsively and mechanically for several seconds before they could emit any sound. At last, in obedience to her spiritual father's command, the nun began to speak, but her voice, which had not spoken for twenty years, sounded hollow and unnatural.

"I am . . . Edith," she said with difficulty. "I am the bethrothed of the dead King Harold."

At the sound of that name, accursed by all pious servitors of the Church, the timid nuns, aghast, made the sign of the cross. But the priest said, "My daughter, you have loved a great sinner on earth. King Harold is accursed by the Catholic Church, the Holy Mother of us all. For him there will never be any forgiveness; he must burn forever in the fires of hell. But God has seen your long, selfless asceticism. He prizes your repentance and your tears. Go in peace. In the domain of Paradise another immortal bridegroom awaits you."

The cavernous, waxen cheeks of the dying woman suddenly colored. In her eyes, so long faded, a passionate, feverish fire flared up.

"I need no Paradise without Harold!" she exlaimed to the horror of all the nuns. "If Harold is not forgiven, then let God not call me to His abode!"

The nuns stood hushed and stunned with dread. Edith, raising herself up from her death bed through superhuman effort, threw herself prostrate at the crucifix.

O great God!" she called out in her broken and hardly human voice. "For one moment of Your Son's suffering, You removed the seal of original sin from all mankind. And I for twenty years have been dying a slow, tormenting death every day, every hour. You saw my suffering, You know it. If I have merited it before You, then forgive Harold! Give me a sign before I die. When we read the Lord's Prayer, may the candle in front of the crucifix light up of its own accord. Then I shall know that Harold is forgiven."

The priest recited the Lord's Prayer. Solemnly, distinctly, he enunciated every word. In a whisper the nuns both young and old repeated the sacred prayer after him. There was none among them not pierced by pity for the unhappy Edith, who would not have given her own life gladly to save Harold's soul.

Edith lay prone on the ground. Her body had already been seized by a convulsion, and all her life as it ran out was concentrated in her eyes alone, fixed upon the crucifix.

But the candle did not light.

The priest finished the recital of the prayer. In a sorrowful voice he pronounced the amen. The miracle had not happened. Harold was not forgiven. From the lips of pious Edith burst forth a howl of curses, and her gaze was extinguished forever.

And this was the novel which caused a crisis in my sister Anyuta's inner life. For the first time, certain basic questions vividly touched her imagination. Is there an after-life? Does everything end with death? Will two beings who love meet in the next world, and will they recognize one another?

With the extremism she brought to everything she did, she was pierced through and through by these questions, as though she were the first to discover them, and it truly began to seem to her that she could not live without an answer.

It was a marvelous summer evening, I remember. The sun had already begun to set; the heat of the day had subsided, and the air was extraordinarily clear and pure. The scent of roses and new-mown hay floated in through the open windows. The lowing of cows, the bleating of sheep, the voices of the workmen carried from the farm—all the myriad sounds of a pastoral summer evening, but so softened and transformed by distance that their harmonious totality only heightened the sense of quietness and peace.

My mood was somehow especially light and cheerful. I had managed to tear myself away for a moment from the governess's vigilant surveillance and had shot like an arrow straight up the staircase to the tower, to see what my sister was doing. And what did I find?

She was lying on the couch with her hair let down, all lit by the rays of the setting sun, and she was sobbing uncontrollably, sobbing so hard that it seemed her breast would burst.

I was horribly alarmed and ran to her.

"Anyutochka, what's the matter with you?"

But she did not answer me, only waved her hand for me to go away and leave her alone. Of course I only insisted harder. For a long time she didn't say anything but at last, raising herself up, she said in a weak and, as it seemed to me, broken voice, "You wouldn't understand it anyway. It's not for myself I'm crying, but for all of us. You're still a child, you don't have to think about serious things. I used to be like that too. But this marvelous, cruel book"—and she pointed to Bulwer-Lytton's novel— "made me look deeper into the mystery of life. And then I understood how illusory everything is that we strive for. The greatest happiness, the deepest love—all of it ends with death. And what waits for us then—and whether anything waits for us—we don't know, and we'll never know, never! Oh, it's horrible! Horrible!" Again she burst into tears and buried her head in the sofa pillow.

This genuine despair of a sixteen-year-old girl sharply brought up against the concept of death by reading a romantic English novel, those bookish, highly emotional words addressed to her ten-year-old sister, all this would doubtless have made an adult smile. But as for me, my heart literally froze with horror. I was filled from head to toe with reverence for the importance and the profundity of the ideas with which Anyuta was engaged. All the beauty of that summer evening darkened for me suddenly, and I even felt ashamed of the inexplicable happiness that had filled my whole being a moment before.

"But after all, we know there is a God, and we'll go up to Him after we die," I tried to object. My sister gave me a benign look, the way an adult looks at a child.

"Yes, you still preserve your pure childish faith. We won't talk about it any more," she said in a voice that was very sad and at the same time filled with such awareness of her superiority over me that for some reason her words made me feel ashamed all over again.

A great change took place in my sister after that evening. For a few days she went about gentle and sad, portraying in her whole mien her renunciation of earthly pleasures. Everything about her said, *memento mori.* The knights with their fair damsels and their love-jousts were forgotten. To what end is love, to what end is desire, when all of it ends in death?

She no longer picked up a single English novel: she had grown sick and tried of them. On the other hand, she eagerly consumed *L'Imitation de Jesus Christ* and resolved, in the spirit of Thomas á Kempis, to muffle the doubts arising in her heart through self-discipline and self-denial. She grew unprecedentedly gentle and gracious with the servants. If I or our younger brother asked her for anything she didn't snap at us the way she sometimes used to do, but yielded immediately, though with an air of such soul-crushing resignation that my heart pinched and all desire for merriment disappeared.

Everyone in the house was filled with respect for her pious mood and treated her tenderly and carefully, as though she were an invalid or a person who had suffered a great grief. Only the governess shrugged her shoulders skeptically, and Papa would tease her at dinner time for her misty expression, *son air ténébreux.* But Anyuta bore Father's mockery submissively and treated the governess with such over-refined courtesy that the latter was, if anything, more infuriated than she would have been by rudeness. Seeing my sister like this, I was not able to feel happy either. I was even ashamed of being insufficiently grieved, and I secretly envied the intensity and depth of her feelings.

This mood of hers did not last very long, however. September the fifth was approaching, my mother's name day, a day which was always celebrated with special solemnity in our family. All the neighbors for fifty versts around used to come to us. Up to a hundred guests would gather, and we always organized something special for that day: fireworks, *tableaux vivants* or home theatricals. The preparations naturally began far in advance.

My mother was a great lover of amateur theatricals, and she herself acted well and with much enthusiasm. That year we had just had a permanent stage built, quite a proper one with wings, a curtain and scenery. There were several confirmed theater-lovers in our neighborhood who could always be enlisted as actors. My mother wanted a home production very much. But now that she had a grown-up daughter she felt a little embarrassed, as it were, to show too much enthusiasm in this matter. She would have liked it to appear that all this had been arranged for Anyuta's pleasure. And just at this point, as if by design, Anyuta had to go and embrace a monastic mood!

I recall how carefully, how timidly my mother approached Anyuta and tried to suggest the idea of a home theatrical. Anyuta did not capitulate immediately. At first she showed great disdain for the whole idea: "Such a fuss—and what for!" At last, she consented, as though deferring to the wishes of the others.

But then the actors came and the process of choosing the play began. This, as one realizes, is no easy matter. The play must be amusing and yet not too free-and-easy, and not requiring a complicated production. We decided on a French vaudeville that year, *Les Oefs de Perette.*

It was the first time that Anyuta was taking part in a home theatrical as an adult, and of course she was given the leading role. The rehearsals began. She revealed an amazing talent for acting. And now the fear of death, the struggle between faith and doubt, the terror of the mysterious beyond—all that evaporated. Anyuta's clear voice trilling French couplets was heard through the house from morning till night.

After Mama's day party Anyuta wept bitterly once again. But now there was a different reason; it was because Father refused to yield to her urgent entreaties to enroll her in a drama school. She felt that it was her life's vocation to become an actress.

CHAPTER NINE

Anyuta's Nihilism

At the time when Anyuta was dreaming of knights and shedding bitter tears over the fate of Edith of the swan's neck and Harold the last of the Saxon kings, most intellectual young people in other parts of Russia were caught up in quite a different current, quite different ideals. Anyuta's enthusiasms, therefore, might seem strangely anachronistic. But that small corner where Palibino was situated was so remote from all the centers, fenced off from the outside world by such thick, high walls, that the wave of new ideas did not reach our quiet cove until long after it had swelled and risen into the open sea.

But for all that, when these new currents reached our shores at long last, they took hold of Anyuta at once and pulled her along with them.

It is difficult to say just how, from where and by what means these new ideas made their way into our household. We know that it is characteristic of every transitional epoch to leave few traces behind. A paleontologist, for example, may study a stratum of some geological cross section and discover in it many fossilized vestiges of clearly differentiated flora and fauna through which he can reconstruct a complete picture of the universe of that period. If he climbs one stratum higher, a totally different formation lies before him, totally new forms—and what their origin is, how they developed from their predecessors, he is unable to say. Fossilized specimens of fully developed forms are to be found everywhere in abundance; the museums are filled with them. But how overjoyed the paleontologist is if somewhere, by accident, he is lucky enough to unearth a skull, a few teeth, a fragment of a single bone of some transitional form through which he can reconstruct in his scientific imagination the course by which the development took place. One might suppose that nature had jealously expunged and smoothed over all the traces of her work. She seems to flaunt the perfected speciments of her creation, those in which she has succeeded in embodying some fully developed idea, but ruthlessly extirpates the very memory of her first uncertain experiments.

The inhabitants of Palibino lived a peaceful and quiet life. They grew and aged, quarreled with one another and then made peace. In order to pass the time, they would dicusss some magazine article on the subject of this or that scientific discovery. They were quite certain, however, that all these matters belonged to a world alien and remote from themselves, which would never impinge directly upon their everyday existence.

And now suddenly, from out of the blue, signs of some strange ferment had appeared, approached closer and closer, no doubt about it, and threatened to undermine the very structure of their quiet, patriarchal way of life. And not from one direction only did the danger threaten; it seemed to come from all sides at once.

It might be said that during the decade between the beginning of the 1860s and the beginning of the 1870s, all the intellectual strata of Russian society were concerned

Sofya's parents, General Vasily Vasilievich and Elizaveta Fyodorovna Korvin-Krukovsky, in 1874

Miss Margaret Smith,
Sofya's English governess

Sofya's tutor, Yosif Malevich

Sofya's mother, Elizaveta Fyodorovna Korvin-Krukovskaya (from a watercolor by L. Bryullov in the collection of the Russian Museum in Leningrad)

Sofya at the age of fifteen

Sofya Kovalevskaya in the 1870s

with a single question: the family discord between the old and the young. No matter which gentry family one might inquire about at that period, one heard the same story: the parents had fallen out with their children. And these quarrels arose not from weighty material causes, but from questions of a purely theoretical and abstract nature.

"They couldn't see eye to eye in their convictions!" That was all; but this "all" was quite sufficient to cause children to leave their parents and parents to disown their children.

The children, particularly the young girls, were seized at that time by something like an epidemic of running away from the parental home. In our immediate neighborhood the Lord spared us, everything was still in order, but rumors were already flying from other parts: now one, now another landowner's daughter had run away—this one to Europe to study, that one to Petersburg, to "the Nihilists."

The chief bugbear of all the parents and mentors in the Palibino district was a certain mythical commune which, according to the rumors, had been established somewhere in Petersburg. It had recruited (or so it was believed, at least) all the young girls who wanted to flee the parental nest. Young people of both sexes lived there in a state of total communism. No servents were allowed, and young ladies of the nobility and the gentry scrubbed floors and polished samovars with their own hands. It goes without saying that none of the persons disseminating these rumors had ever actually been inside this commune. Where it was located and how it could exist in Petersburg at all, under the very nose of the police, no one knew exactly. Nevertheless, no one had the slightest doubt that such a commune existed.[1]

Soon the signs of the times began to be noticeable in the immediate vicinity of our house. Our parish priest, Father Filipp, had a son who had hitherto gladdened his parents' hearts with his obedience and his good behavior. Suddenly, after completing his course at the Seminary practically first in his class, this exemplary youth, for no discoverable reason, turned into a rebellious son and refused pointblank to enter the priesthood, although all he had to do was stretch out his hand to be given a remunerative parish.

The Right Reverend Bishop himself summoned the young man and tried to dissuade him from leaving the bosom of the church, hinting clearly that all he had to do was to express the wish to be appointed parish priest of the village of Ivanovo, one of the richest in the province. Of course he would first have to marry one of the daughters of the former priest, for it was a hallowed tradition that the parish should be part of the dowry, as it were, of a daughter of the late priest. But even this enticing prospect failed to tempt the young priest's son. He preferred to go off to Petersburg to enter the university as a paying student and dedicate himself to four years of study on tea and a crust of dry bread.

Poor Father Filipp grieved over his son's folly. But he might still have felt consoled if only the son had matriculated in the Faculty of Law which, as is well known, is the most lucrative. Instead of doing that, however, the boy went into the natural sciences. And during his very first holiday he came home and started talking rubbish to the effect that man, allegedly, is descended from monkeys, and that Professor Sechenov had proved, allegedly, that there is no soul, only reflexes. Poor Father Filipp was so taken aback that he grabbed his holy water basin and started to sprinkle his son from it.

In former years when the young man came home on holiday from the Seminary he had never failed to attend a single one of our family's celebrations, but invariably appeared to pay his respects. At the holiday feast he would sit at the lower end of the table as befits a young man of his station, eating the name-day cake with gusto, but not mixing into the conversation.

This summer, however, the young man was distinguished

by his absence from the very first name-day celebrated after his arrival. But to make up for it, he showed up at our house on an inappropriate day, and when he was asked by the servant what he wanted, he replied that he had simply come to pay the General a visit.

My father had already heard quite a good deal of talk about the priest's "Nihilist" son. He had not failed to observe the boy's absence from his name-day party, although, of course, he gave no sign of having taken notice of such a trivial circumstance. But now he was outraged at the idea that the young upstart had taken it into his head to show up uninvited, as an equal, and he resolved to teach him a good lesson. Therefore, he had the boy informed by his lackey, Ilya, that "the General receives people who come to him on business, and petitioners in the mornings only, up to one o'clock."

Faithful Ilya, always quick to scent out the master's intentions, fulfilled his mission in precisely the spirit in which it had been conveyed to him. But the young priestling was not the least bit disconcerted. As he was leaving he remarked in a very free-and-easy manner, "Tell your master I won't be setting foot in this house any more."

Ilya fulfilled this commission as well, and it can be imagined what a furore the young man's effrontery raised, not only in our house but throughout the district. But what was even more striking was the fact that Anyuta, when she heard about the incident, ran to Father's study on her own initiative. With flaming cheeks, panting with excitement, she blurted out, "Why did you have to go and hurt Alexei Filippovich's feelings, Papa? That's terrible—it's not worthy of you to insult a decent person like that."

Father stared at her, amazed. His surprise was so great that for a few seconds he couldn't even find the words to respond to the insolent wench. But Anyuta's sudden attack of daring had already evaporated, and she hastened back to her own room.

Once recovered from his astonishment and after thinking things through, Father decided that it would be better not to make too much of his daughter's impudence but to treat the whole thing as a joke. At dinner in Anyuta's presence he told a story about a certain princess who took it into her head to intercede for a stable boy. Both the princess and her protégé, of course, were put into a devastatingly comical light. Father was a master at witticisms, and we were all terribly afraid of his gibes.

This time, however, Anyuta listened to his tale without the slightest embarrassment—on the contrary, with a saucy and defiant expression on her face. She then went even further and expressed her protest against the insult to which the priest's son had been subjected by seeking out every possible pretext for meeting him, either at some neighbor's house or while out walking.

One day at supper in the servant's hall Stepan, the coachman, told how he had seen the elder young lady with his own eyes, strolling in the forest alone with the priest's son.

"And wasn't it fun just to watch them! The young mistress is walking along not saying a word, with her eyes cast down, and she's playing with her parasol. And he's stepping right along next to her with those big, long legs of his, looking just like a long, gawky stork. And he's talking his head off the whole time and waving his hands around. And then all of a sudden he pulls some shabby old book out of his pocket and starts reading it out loud, like he's giving her a lesson!"

It should be stated here that the young man actually bore very little resemblance to that storied prince or medieval knight Anyuta had once dreamed about. His ungainly, lanky body, long veiny neck and pale face fringed with thin rusty-blond hair, his big red hands with their flat and not always irreproachably clean fingernails; and worst of all, his unpleasant, vulgar accent overemphasizing the "o's" and bearing unmistakable witness to his clerical origins and Seminary education—none of this made him a very attractive hero in the eyes of a young girl of aristocratic tastes and habits. It would have been hard to suspect that Anyuta's interest in the priest's son had any romantic basis. There was obviously something else involved.

And, as a matter of fact, the young man's chief glamor for Anyuta consisted in the fact that he had just arrived from Peterburg and brought with him the very latest and newest ideas. As if that weren't enough, he had even had the good fortune to see with his own eyes (granted, from a distance only) many of those great men[2] who were worshipped by all the young people of that period.

This was quite enough to make him extremely interesting and attractive. But on top of everything else, it was thanks to him that Anyuta could get hold of various books otherwise inaccessible to her. At home we received only the most moderate and staid periodicals: of the foreign ones, the *Revue des Deux Mondes* and *Atheneum,* and of the domestic, only *The Russian Messenger.* As a great concession to the spirit of the times, my father had consented to subscribe that year to Dostoevsky's journal *The Epoch.*[3] But through the priest's son Anyuta could obtain periodicals of quite a different stripe: *The Contemporary, The Russian Word,* each new issue of which was regarded as the event of the day by the young people of that time. On one occasion he brought her an issue of Alexander Herzen's illegal publication *The Bell.*[4]

It would not be accurate to say that Anyuta immediately and uncritically accepted all the new ideas propagandized by her friend. Many of them made her angry, seemed too extreme; she resisted and argued. At all events, however, she developed very rapidly under the influence of her discussions with the young man and her reading of the books he brought her, and she changed with each passing day.

Toward autumn the priest's son had such a basic disagreement with his father that the latter told him to leave the house and not come back on his next vacation. But the seeds planted in Anyuta's mind went on growing and developing.

She changed even outwardly. She began to dress very simply in black dresses with plain collars, to wear her hair pulled back and covered by a net. She spoke contemptuously now of balls and social visits. In the mornings she would call the servant's children together and teach them to read, and when she met peasant women out walking, she would stop them and have long conversations with them.

But the most remarkable thing of all was that Anyuta, who had formerly loathed her studies, now displayed a passion for learning. Instead of spending her pocket money on frills and furbelows as she used to do, she would now order whole boxes of books—and not novels, mind, but books bearing such sage titles as *The Physiology of Life, The History of Civilization* and so forth.

One day she went to my father and expressed a totally unexpected demand: that he should allow her to go to Petersburg alone, to study. At first he tried to turn her request into a joke, his old way of dealing with this sort of thing when Anyuta used to announce that she didn't want to go on living in the country. But this time Anyuta could not be stopped. Neither jokes nor gibes had any effect on her. She argued heatedly that the fact that Father had to live on his estate did not mean that she, too, must be shut away in the country where there was neither fun nor serious work.

Finally Father lost all patience and shouted at her as if at a small child.

"If you yourself don't understand that it is the duty of every respectable young woman to live with her parents until she gets married, then I'm not about to argue with a stupid brat!"

Anyuta realized that it was futile to go on insisting. But from that day the relationship between her and Father grew very strained. They showed a mutual irritation with one another, and this irritation increased with each day. At dinner, the only time of the day when they met, they hardly ever addressed one another directly—but a dig or a caustic hint could be sensed in every word they said.

And in general, an unheard-of discord now began to develop in our family. Even before that we had had few interests in common, but then all the members of the family had kept to themselves, taking scant notice of one another. But now, two hostile camps seemed to have formed.

The governess had come out violently against all the new ideas from the very beginning. She dubbed Anyuta "the Nihilist" and "the progressive young lady." This latter label, somehow, sounded especially biting from her lips. Intuitively sensing that Anyuta had some special project in her mind, she started to suspect her of the most criminal intentions: running away from home on the sly, marrying the priest's son, joining the notorious commune. Therefore, vigilantly and mistrustfully, she began to keep track of Anyuta's every step. And Anyuta, feeling that the governess was keeping her under surveillance, began deliberately, for the purpose of annoying her, to cloak her behavior in exacerbating and offensive mysteriousness.

The militant atmosphere which now prevailed at home was quick to react upon me as well. The governess, who even before had disapproved of my closeness to Anyuta, now insulated her charge from the "progressive young lady" as if from an infectious disease. She prevented me from being alone with my sister as much as she could, and regarded as some kind of criminal act every attempt of mine to run out of the classroom and upstairs to the world of the grownups.

I was extremely annoyed with the governess's watchful supervision. I, too, sensed intuitively that Anyuta had developed new and unprecedented interests, and I wanted terribly to understand what it was all about. Almost every time I ran into her room unannounced, I would find her sitting at her desk and writing something. A few times I tried to find out what it was that she was writing. But since she had been reprimanded by the governess more than once, not only for straying off the straight path herself, but also for desiring to corrupt her sister as well, she was afraid of fresh rebukes and never failed to chase me away.

"Oh, please get out of here," she would say impatiently. "If Margarita Frantsevna finds you here—we'll both catch it then!"

I would return to the classroom feeling vexed and resentful of the governess, on whose account my sister didn't want to tell me anything. It grew more and more difficult for the poor Englishwoman to get along with her charge. What I understood from the discussions carried on at the dinner table was mainly that obedience to one's elders was no longer in fashion. Consequently, the feeling of subordination in me weakened noticeably.

There was now a quarrel between the governess and myself almost every day, and after one particularly stormy scene, she announced that she could not remain with us any longer.

Inasmuch as leaving us was something she had previously threatened to do more than once, I paid little attention to it at first. But this time the matter turned out to be more serious. On the one hand, the governess had already gone too far and could not honorably retreat from her threat. On the other, the continual scenes and upsets had so strained everyone's nerves that my parents, in the hope that peace might return to our household if she left, did not try to keep her.

But up to the very last minute I did not believe that the governess was really going, not until the actual day of her departure.

CHAPTER TEN

Anyuta's First Literary Experiments

A big, old-fashioned suitcase, neatly encased in a canvas cover and tied with rope, had been standing in the front hall since morning. Over it towered a whole battery of little boxes, baskets, bags and bundles, the sort of thing no old maid can travel without. The old springless carriage, hitched up to a troika of horses with the oldest and least elegant harness, the one Yakov, the coachman, always used when a long journey was in the offing, was already waiting at the entrance. The maids were fussing about, bringing different trifles and trinkets in and out, but Papa's valet, Ilya, stood motionless, leaning languidly against the door jamb and expressing through his whole contemptuous posture the opinion that the impending departure was of no moment, not worth raising a household commotion over.

Our whole family had gathered together in the dining room. Following tradition, Father invited everyone to sit down before the journey. The masters occupied the front corner while the full assembled body of domestics crowded together at some distance apart, perched respectfully on the edges of their chairs.

A few minutes passed in reverential silence during which the spirit was involuntarily gripped by nervous anguish, that sensation inevitably evoked by every departure and every separation. But now Father gave the signal to rise and crossed himself before the icon. The rest followed his example, and then the tears and the embraces began.

Now I looked at my governess wearing her dark traveling dress and wrapped in a warm down shawl, and suddenly she appeared totally different from the way I was used to seeing her. She seemed to have aged all at once: her full, vigorous body looked pinched, those "lightning-bolt" eyes of hers, as we furtively and jeeringly used to refer to them—those eyes which no shortcoming of mine ever escaped—were red now, swollen and brimming with tears. The corners of her mouth twitched nervously. I saw her as pitiful for the first time in my life.

She held me in her arms long and convulsively, with such fierce tenderness as I never would have expected of her.

"Don't forget me. Write to me. It's not so easy to part with the child I brought up from the age of five," she said with a sob. I pressed myself against her, too, and began to weep desperately. I was overwhelmed by piercing anguish, by a sense of irretrievable loss, as if our whole family were disintegrating with her departure. And mixed with this was an awareness of my own guilt as well. I was painfully ashamed to recall that all through these past few days, as recently as this very morning, I had been secretly gloating at the thought of her leaving and the freedom in store for me.

"And now I've got what I wanted, now she's really going, now we'll be without her!" And I felt so badly for her at that moment that I was ready to give goodness knows what to keep her. I clung to her as if I could not tear myself away.

"Time to go now, if she's to reach town before dark," someone said. The luggage was already stowed away inside the carriage. The governess was helped in. One more long, tender hug.

"Careful now, young lady, don't fall under the horses!" someone shouted, and the carriage started to move.

I ran upstairs to the corner room from whose windows could be seen the verst-long, birch-lined avenue leading from the house to the high road, and I pressed my face against the window pane. As long as the carriage was visible I could not tear myself away from the window, and the feeling of my own guilt grew stronger and stronger. Lord, how sorry I felt for the departed governess! All my clashes with her—and they had been especially frequent of late—appeared to me now in an entirely different light from before.

"But she loved me. She would have stayed if she only knew how much I loved her. And nobody loves me now, nobody!" I thought, in belated regret, and my sobbing grew louder and louder.

"Is that Margarita all this grieving is about?" my brother Fedya asked as he ran by. Surprise and mockery were in his voice.

"Leave her alone, Fedya, it does her honor that she's so affectionate," I heard my old aunt's didactic voice behind my back. None of the children liked her—they regarded her as insincere, for some reason. Her saccharine praise had the same unpleasant and sobering effect on me as my brother's gibes. Ever since I was a little girl, I had not been able to bear attempts at consolation from people I didn't care for. So I angrily pushed away the hand that my aunt, trying to caress me, had placed on my shoulder, and I ran to my own room muttering irately, "I'm not grieving at all and I'm not affectionate at all."

The sight of the deserted classroom almost awakened a new paroxysm of despair. Only the realization that nobody would hinder me anymore from being with my sister as much as I liked comforted me a little. I decided to run to her right away and see what she was doing.

Anyuta was pacing back and forth in the big salon. She always took a "constitutional" when she was particularly preoccupied or troubled. Her expression at such times was abstracted, her luminous green eyes grew quite transparent and saw nothing of what was going on around her. Without being aware of it herself, she kept time with her own thoughts. If they were melancholy ones, her gait grew languid and slow also, but if her thoughts livened up and ideas started coming to her, her walk would speed up so that at last she was no longer walking but running around the room.

Everyone in the house was familiar with this habit and teased her about it. Often I would stealthily watch her walking, and I would have loved to know what Anyuta was thinking about. I knew from experience that it was no use bothering her at times like these. And yet, seeing that the pacing went on and on, I finally lost my patience and made an attempt to start talking.

"Anyuta, I'm so bored! Give me one of your books to read!"

But she kept on pacing as though she hadn't heard me.

Again, a few minutes of silence.

"Anyuta, what are you thinking about?"

"Oh, go away!" was her contemptuous reply. "You're too little for me to tell you everything."

By this time I was highly insulted. "So that's what you're like! You don't even want to talk to me! And here I thought you and I would be friends now that Miss Smith's gone, and you chase me away! All right for you then, I'm going away and I'll never love you any more, not one bit!"

I was almost in tears and on the point of leaving, but my sister called me back. The truth was that she herself was burning with the desire to tell someone what it was that she was so engrossed with. But since there was no one else in the family she could tell it to, even a twelve-year-old sister would have to do for lack of a better audience.

"Listen here," she said. "If you promise me never to say a word to anybody, that you won't go blabbing no matter what . . . I'll trust you with a big secret."

My tears dried up in a twinkling. The anger was quite gone. Of course I swore that I would be as silent as a clam, and I waited impatiently to hear what she was going to tell me.

"Let's go up to my room," she said solemnly. "I'll show you something . . . something . . . that you probably never expected."

And now she led the way to her room and over to the old chest of drawers in which, I knew, she kept her most cherished secrets. Slowly, without hurry, so as to prolong the suspense, she unlocked one of the drawers and took out of it a large businesslike envelope with a red seal on which was engraved: *The Epoch.* It was addressed to Domnya Nikitishna Kuzmina, our housekeeper, who was utterly devoted to my sister and ready to go through fire and water for her. From this envelope my sister drew out another envelope, a little smaller, on which was marked, "For delivery to Anna Vasilievna Korvin-Krukovskaya." And finally she handed me a letter closely written over in a bold masculine hand.

I do not have that letter now, but I read and reread it so often in my childhood, and it etched itself so deeply in my memory, that I think I can repeat it almost verbatim.[1]

> "Dear Madam, Anna Vasilievna! Your letter, full of sincere and gracious confidence in me, interested me so much that I began at once to read the story you enclosed with it.
>
> "I confess that I began reading not without a secret trepidation. So often we magazine editors are faced with the melancholy duty of disillusioning the young

novice authors who send us their first literary efforts for evaluation. In your case, this would have been most regrettable to me. But as I read my fear was dispelled, and I yielded more and more to the charm of that youthful directness, sincerity and warmth of feeling which permeate your story.

"These qualities so won me over that I fear I may still be under their influence. Therefore I do not dare to give you a categorical and objective answer to the question you ask—will you develop with time into a major writer?

"One thing I can say: your story will be published by me (and with great pleasure) in the next issue of my journal.[2] As for your question, I advise you to go on writing and working. The rest, time will tell.

"I shall not conceal from you the fact that your story has much in it that is still unfinished, over-naive. There are even—foregive my candor—errors in Russian grammar. But all these are minor flaws which you can overcome by hard work; the general impression remains most favorable.

"Therefore, I repeat: write and keep writing. I would be sincerely happy if you would tell me some more about yourself: how old you are, what circumstances you live in. It is important for me to know all this in order to assess your talent correctly."

"Devotedly yours, FYODOR DOSTOEVSKY."[3]

I read this letter, and in my amazement the lines swam before my eyes. I knew the name of Dostoevsky; lately it had come up often at our dinner table during my sister's arguments with my father. I knew that he was one of the most distinguished Russian writers. But why in the world was he writing letters to Anyuta, and what did it all mean? For a moment, it occurred to me that my sister might be making a fool of me so as to laugh at me later for my credulousness.

After I had finished reading the letter I looked at Anyuta in silence, not knowing what to say. She was obviously delighted with my bewilderment.

"Do you understand it? Do you understand?" she said finally, in a voice breaking in happy excitement. "I wrote a story and sent it to Dostoevsky without saying a word to anyone about it. And now—see—he thinks it's good—and he's going to publish it in his magazine! So my great dream has come true! I'm now a Russian authoress!" she cried out in a surge of uncontrollable joy.

To understand what the word "authoress" meant to us, one must bear in mind that we lived in a rural backwater removed from any hint, even a feeble hint, of literary life. A good deal of reading went on in our family, and many books were ordered. Not only we, but everyone else in our circle responded to every book, to every printed word as if it had come across vast distances, from some unfathomed and alien world which had nothing in common with us.

Strange as it may seem, it was a fact that up to then neither my sister nor I had ever even seen anyone who had published a single line. True, there was a certain teacher in our provincial town of whom it was rumored that he had written a dispatch about our district for the newspapers. I remember the respectful awe with which everyone behaved to him until it came out later that he wasn't the one who had written the dispatch at all—it was some passing journalist from Petersburg.

And now, all of a sudden, my sister—an authoress! I couldn't find words to express my delight and astonishment to her. All I could do was throw myself on her neck, and for a long time we petted one another, and laughed, and said all kinds of foolish things in our happiness.

Sister could not bring herself to discuss her triumph with anyone else at home. She knew that everyone, even our mother, would get scared and tell my father about it. And in his eyes her act—writing an unsolicited letter to Dostoevsky and subjecting herself to his judgment and derision thereby—would have seemed a dreadful crime.

Poor Father! How he loathed female writers, and how he suspected every last one of them of behavior which had nothing in common with literature! And it had to be his fate to become the father of an authoress.[4]

His personal acquaintance with women writers was limited to one, the Countess Rostopchina.[5] He had seen her in Moscow at a time when she was a brilliant society beauty for whom all the young noblemen of the period, my father included, sighed hopelessly. Then, many years later, he met her abroad somewhere, I believe in Baden-Baden, in the roulette hall.

"I looked at her and couldn't believe my eyes," Father often recounted. "The Countess was walking along followed by a whole trail of rogues, each one grosser and more vulgar than the next. The were all yelling, laughing, guffawing, treating her very familiarly. She walked up to the game table and started tossing on one gold piece after another. Her eyes were burning, her face was all red, her chignon was pushed askew. She went and lost it all, right down to the last gold piece, and then she shouted to her aides-de-camp, 'Eh bien, messieurs, I'm cleaned out now. No more stakes. Let's go drown our sorrows in champagne!' And that's what happens when a woman starts writing books!"

It is understandable, therefore, that my sister was in no hurry to go boasting to him of her success. But the mystery in which she had to cloak her debut in the field of literature endowed it with a special charm. I remember how thrilled we were when an issue of *The Epoch* arrived a few weeks later and we saw, listed in the table of contents; *"The Dream," a story by Yu. Orbelov* (Orbelov was the pseudonym Anyuta had selected, since she obviously could not publish under her own name).

Naturally, she had already read me the story from the rough copy she had kept. But now, printed in the pages of the journal, it seemed to me brand-new and amazingly beautiful.

Its contents were as follows: the heroine, Lilenka, lives among middle-aged people worn down by life, who hide away in a quiet corner seeking peace and forgetfulness. They strive to inculcate in Lilenka their own fear of life and its disturbances. But she is lured and attracted by this unknown life, whose faint echoes reach her only as the remote plash of a wave hidden beyond the mountains of the sea. She believes that there are places

> Where people live more gaily,
> Where they live life and don't weave spiderwebs.

But how to get to people like that? Unnoticed even by herself, Lilenka has been infected by the prejudices of the milieu in which she lives. Almost unconsciously she is faced with a question at every step: is this or is this not proper behavior for a well-born young lady? She would like to break out of her constricted world, but she is frightened by everything vulgar, everything not *comme il faut.*

One day during a stroll in the city she meets a young student (it goes without saying that the hero of that period had to be a student). This young man impresses her deeply, but after the fashion of a well-bred young lady she does not show that she likes him. And so the acquaintance breaks off with this meeting.

For a time after that Lilenka pines for him but then calms down. Only when, amid the different mementos of her colorless existence which fill her bureau drawers as is the case with most young ladies, she chances to come upon a knickknack that reminds her of that unforgettable evening, she hastily bangs the drawer shut and goes about dissatisfied and sullen all the rest of the day.

But then one night she has a dream. The student comes to her and reproaches her for not following him. In her sleep a series of scenes succeed one another: an honest, hard-working life with the person she loves in a circle of intelligent friends, a life filled with warm, serene happiness in the present and an immeasurable store of hope for the future. "Look and repent!" the student tells her. "This is what our life together would have been." And then he disappears.

She wakes up. Under the spell of her dream she resolves to throw aside her worry about proper behavior for a young lady. She, who until then had never left her house unaccompanied by a maid or lackey, now goes away secretly. She takes the first available cabby and drives to the faroff, shabby street where—she knows—her dear student lives. After many searches and adventures caused by her inexperience and impracticality, she locates his apartment at last.

But once there, she finds out from his roommate that the unfortunate young man had died of typhoid fever a few days before. The friend tells her how hard the student's life was, what bitter need he endured and how, in his delirium, he had repeated the name of a young lady. In consolation or reproach of the weeping Lilenka, he recites Dobrolyubov's stanzas.

> I fear death too may play
> A scurvy joke on me.
> . . . .
> I fear that all I craved in life
> So greedily, so futilely,
> May smile consolingly at me
> Above my tomb.[6]

And so Lilenka returns home. None of her family ever learns where she disappeared that day. But she herself retains forever the conviction that she has thrown her happiness away. She does not survive for long and dies bewailing her vainly wasted and empty youth.

This was the content of my sister's first story. Its chief merit was not the plot at all, but the aliveness, the reality with which she was able to convey her heroine's transports of feeling. She had experienced them during her own pacing back and forth. She was particularly successful also in her depiction of the scenes of happiness fantasied by Lilenka. She had pictured these scenes so distinctly for her own self; she longed so for that happiness, she believed so deeply in the possibility of it.[7]

Anyuta's first success encouraged her very much and she immediately started a second story, which she finished within a few weeks. This time the hero of her tale was a young man, Mikhail, brought up in a monastery far from his family by his uncle, a monk. Dostoevsky liked the second story much more than the first and considered it more mature.[8]

The character Mikhail shows a certain similarity to Dostoevsky's Alyosha in *The Brothers Karamazov.* When I read that novel some years later as it appeared in magazine installments, this similarity caught my eye and I remarked on it to Dostoevsky, whom I was seeing very often at that time.

"You may be right at that!" said Fyodor Mikhailovich, striking his forehead with his hand. "Believe it—I really forgot Mikhail when I was working out my Alyosha. Although," he added, after a short pause, "it may be that it came to me unconsciously."

The publication of Anyuta's second tale, however, did not turn out as happily as the first. By lamentable mischance Dostoevsky's letter came into Father's hands, and a terrible scene took place.

Again it happened on the fifth of September, a memorable day in the annals of our family. As usual, we had many guests. It happened to be the day when we received our mail, which was delivered to our estate only once a week. Normally the housekeeper, whose name Anyuta was using for her correspondence with Dostoevsky, went out to meet the postman and take her letters away before he brought the mail in to the master. But this time the housekeeper was busy looking after the guests.

Unfortunately, the postman who usually delivered our mail had taken a drop too much on the occasion of the mistress's name day: in other words, he was dead drunk. In his place they sent a young fellow who didn't know the established procedure. And so the mailbag went directly to Father's study without its preliminary inspection and weeding out.

My father's eye fell at once upon a registered letter with the seal of the journal *The Epoch,* addressed to our housekeeper. What kind of strange business was this? He summoned the housekeeper and made her open the letter in his presence. One can, or perhaps one cannot, imagine the scene which followed.

As bad luck would have it, that same letter contained Dostoevsky's payment to my sister for her two stories: three hundred and some odd rubles, as I recall. This circumstance—the fact that Anyuta was clandestinely receiving money from a strange man—seemed to my father so disgraceful and scandalous that he fell ill. He had a chronic heart condition and gallstones as well, and the doctors said that any excitement was dangerous for him and might lead to sudden death. The possibility of such a disaster was a common nightmare for all the members of our family. Whenever the children caused him any unpleasantness his face would turn dark, and we would be overwhelmed with dread that we might kill him. And now—such a shock! And, as if by design, the house was filled with guests!

A regiment was quartered in our provincial town that year. On the occasion of Mama's name-day, all the officers together with their colonel had come to us and brought with them the regimental band as a surprise.

The name-day banquet had finished about three hours earlier. All the chandeliers and candelabras were lit in the great hall on the second floor and the guests, who had already taken their postprandial rests and changed into their ball clothes, were beginning to assemble. Puffing and preening, the young officers were pulling on their white gloves. The young ladies, in their light-as-air tarlatan dresses with the huge crinolines then in fashion, were admiring themselves before the mirrors.

My Anyuta usually adopted an arrogant attitude to all such people, but now the elegant setting, the dance music, the blazing illumination, her awareness that she was the most beautiful and exquisitely dressed girl at the ball—all this intoxicated her. Forgetting her new dignity as a Russian authoress, forgetting how remote was the resemblance of these red-faced, perspiring young officers to those ideal human beings she dreamed about, she twirled among them, smiling at each and every one and basking in the consciousness that she was turning all their heads.

We awaited only my father's appearance for the dancing to begin. Suddenly a servant came into the room, walked up to my mother and informed her, "His Excellency is not feeling well. He asks you please to come to his study."

We felt very uneasy. Mama hurriedly got up, lifted the train of her heavy silk gown with her hand, and left the hall. The musicians, who were in the next room waiting for the agreed-upon signal to start the quadrille, were ordered to wait.

A half hour went by. The guests began to worry. At last Mama returned. Her face was red and agitated, but she made an effort to appear calm and smiled constrainedly. To our guests' solicitous question, "What's the matter with the General?" she answered evasively, "Vasily Vasilievich is not feeling quite well and begs you to excuse him and begin the dancing without him."

Everyone realized that something was wrong, but out of politeness they did not press her further. Moreover, they all wanted to start dancing as soon as possible, since they were assembled and dressed up for the occasion. And so the dancing began.

Passing our mother during a figure of the quadrille, Anyuta anxiously looked into her eyes and read in them that something was very wrong. Seizing a moment during the intermission between two dances, she took Mama aside and insisted on being told what was happening.

"What have you done!" said poor Mama, trying hard to hold back her tears. "It's out in the open! Papa read the letter Dostoevsky sent you and nearly died on the spot with shame and mortification."

Anyuta turned white as a sheet, but Mama went on, "Please keep yourself under control, even now. Remember that we have a houseful of guests who would be only too happy to spread gossip about us. Go and dance as if everything were normal." And so Mama and Sister went on dancing almost up to morning, both of them beside themselves with fear at the thought of the storm which would break over their heads the moment the guests had left.

And indeed it was a terrible thunderstorm. As long as the guests were still in our house, Papa allowed no one in to see him and sat locked in his study. During the intermissions between dances my mother and sister kept running out of the ballroom to listen at his door, but did not dare to go inside and returned to their guests tormented by worry about him: How was he feeling now? Was he feeling worse?

When all was quiet in the house, he summoned Anyuta to his study and upbraided her mercilessly. One sentence in particular etched itself very deeply in her memory: "Anything can be expected from a girl who, in secret from her father and mother, is capable of entering into a correspondence with a strange man and taking money from him. Now you are selling your stories, but the time will come—mark my words—when you'll sell yourself."

As she heard these terrible words, poor Anyuta grew cold. She must have known in her heart that they were nonsense. Still, Father spoke with such assurance, in a tone of such profound conviction, his face was so crushed and miserable and, moreover, his authority in her eyes was still so powerful that despite herself, if only for a moment, a tormenting doubt materialized: Had she done a bad thing? Had she committed some disreputable act without being aware of it herself?

As was always the case after every domestic upheaval, everyone in the house had a hangdog look for a few days. The servants learned the whole story immediately. Papa's valet, Ilya, following his commendable habit, had eavesdropped on the entire discussion between my father and my sister and had interpreted it after his fashion. The news of what had occurred, in distorted and exaggerated form, of course, made the rounds of the entire district, and for a long time afterward the neighbors' main topic of conversation was the "horrible" deed committed by the young lady of Palibino.

Little by little, however, the storm subsided. A phenomenon took place in our family which happens frequently in Russian families: the children reeducated the parents. This process of reeducation began with our mother. In the beginning she took his side completely, as always during a clash between father and children. She was terrifed of his falling ill and outraged that Anyuta could upset her father so. But when she saw that her efforts at persuasion were of no avail and that Anyuta went about wounded and disconsolate, she began to feel sorry for her. Soon she was curious to read Anyuta's story, and then there developed a secret pride that her daughter was . . . an authoress. In this way her sympathy shifted over to Anyuta's side, and Father felt himself quite alone.

At first, in the heat of his anger, he damanded that his

daughter promise to stop writing and would agree to forgive her only under this condition. Anyuta, it goes without saying, would not agree to give any such promise. Consequently they did not speak to one another for days on end, and Anyuta did not even appear at dinner.

My mother ran from one to the other, entreating and persuading.[9] Finally, Father capitulated. The first step on the road to conciliation was his agreement to have Anyuta's story read to him.

The reading proceeded in great solemnity. The entire family was present. Fully aware of the importance of the moment, Anyuta read in a voice trembling with excitement. The situation of the heroine, her endeavors to liberate herself from her family, her sufferings under the yoke of the constraints placed upon her—all this was so like the real-life situation of the author that the coincidence was glaringly evident.

Father listened without saying a word all through the reading. But when Anyuta reached the final pages and, barely holding back her own sobs, started reading about Lilenka on her death bed, bewailing her wasted youth, big tears suddenly appeared in his eyes. He got up without a word and walked out of the room. Neither that evening nor in the days following did he say anything to Anyuta about her story. He only addressed her with amazing gentleness and tenderness, and everyone in the family understood that her cause was won.

In truth, an era of mildness and conciliation in our household began from that day forward. The first manifestation of this new era was that our housekeeper, whom Father had dismissed from her job in his first burst of anger, was vouchsafed his gracious forgiveness and remained at her post.

The second conciliatory measure was even more striking. Father gave Anyuta permission to write to Dostoevsky with the single condition that she show him the letters.

Moreover, he promised that on the next trip to Petersburg he would meet Dostoevsky personally.[10]

My mother and sister, as I have already mentioned, made a trip almost every winter to Petersburg, where we had a whole colony of maiden aunts who occupied an entire house on Vasilievsky Island. When my mother and sister came, two or three rooms would be set aside for their use.

Father usually remained in the country. I would be left at home too, in the care of my governess. But that year, inasmuch as the Englishwoman had left and the newly arrived Swiss governess did not as yet enjoy sufficient confidence, my mother decided to take me along with her, to my indescribable joy.

We left in January to utilize the last passable road of winter. The journey to Petersburg was no easy matter. We had to drive approximately sixty versts by dirt road using our own horses, then about two hundred more on the highway by post horse, and finally, about twenty-four hours by train.

We set out in a big covered wagon on runners. In it sat Mama, Anyuta and I, drawn by six horses, while in front of us drove the maid and the luggage, in a sled hitched up to a troika of horses with little bells attached to their harness. All the way we were accompanied and sung to sleep by the resonant voices of the bells, now coming close, now drawing away, now completely hushed in the distance, and then suddenly starting up for a while again, right under our ears.

How many preparations were made for that journey! In the kitchen so many delicious things were cooked and baked that they would have sufficed, it seemed, for an expedition. Our cook was renowned throughout the district for his puff paste, and never did he apply so much effort in this matter as when he baked his butter-dough pies for the master's family, for the road.

And what a marvelous road it was! The first sixty versts led through dense, thick-trunked pine forest intersected only by a multitude of lakes and ponds. In winter these lakes became vast snow fields against which the dark pines were vividly patterned.

It was lovely driving during the day, but even better at night. I would doze off for a moment and then suddenly wake up with a jolt. For the first second I didn't know where I was. At the top of the wagon a tiny lamp gave off the faintest of glimmers, lighting two strange sleeping figures in bulky furs and white travel hoods. I didn't recognize them right away as my mother and sister. Fantastic silvery patterns showed on the frozen glass of the wagon panes. The little bells tinkled without a stop.

All this was so strange, so out of the ordinary that I couldn't figure any of it out at first; I only felt a dull ache in my arms and legs because of my awkward position. Suddenly awareness came to me like a brilliant ray—where we were, where we were going, and how much that was good and new lay in store for us. And my heart overflowed with a vivid thrill of happiness.

Yes, that was a marvelous road! And it has remained perhaps the happiest memory of my childhood.[11]

CHAPTER ELEVEN

Our Friendship With Fyodor Mikhailovich Dostoevsky

As soon as we arrived in Petersburg[1] Anyuta wrote to Dostoevsky and asked him to come and see us. He came on the appointed day. I remember our feverish anticipation as we waited for him, how we listened to every ring in the entrance hall an hour before he was due to arrive. But this first visit of his was a total failure.

My father, as I have already mentioned, was very suspicious of everything that belonged to the literary world. Even though he had given my sister permission to make Dostoevsky's acquaintance, he had done it with many misgivings and not without secret trepidation.

"Remember, Liza, that the responsibility lies on you," was the burden he placed on my mother when he sent us off from Palibino. "Dostoevsky is not a man of our social world. What do we know about him? Only that he's a journalist and a former convict. Fine recommendation, I don't think! You'll have to be extremely careful about him."

Therefore, he gave our mother strict orders to be present without fail when Anyuta met Dostoevsky and not to leave the two of them alone for a minute. I also begged to be present during his visit. Two old German aunties kept contriving pretexts for showing up in the room and gazed at Dostoevsky curiously, as if at some exotic beast. Finally they, too, ended by sitting down on the couch and staying there until the visit was over.

Anyuta was nettled because her first meeting with Dostoevsky, of which she had been dreaming for so long in advance, was taking place under such absurd circumstances. Putting on her angry moue, she remained stubbornly silent. Fyodor Mikhailovich also felt awkward and out of sorts in this tense atmosphere. Amid all these elderly ladies he was both disconcerted and exasperated. He seemed old and ill that day, as he always did when he was in a bad mood. He kept pinching nervously at his sparse light-brown beard and biting at his moustache, and his whole face twitched.

Mama tried with all her might to get an interesting conversation started. Smiling her most social and pleasant smile and yet obviously bashful and awkward, she searched for something agreeable and flattering to say to him and for some intelligent question to pose. Dostoevsky replied in monosyllables, with calculated brusqueness. At last, having exhausted all her resources, Mama too fell silent. Fyodor Mikhailovich sat there for a half hour and then picked up his hat. Awkwardly and hastily he said his goodbyes, and without offering his hand to anyone, he left.

When he was gone Anyuta ran straight to her own room, threw herself down on the bed, and burst into tears.

"Every time! They spoil it every time!" she kept repeating while sobbing convulsively.

Poor Mama felt herself unjustly blamed, and her feelings were hurt that for all her efforts to please everybody, they were all angry with her. She too started to cry.

"You're always like that—never satisfied. Father did it your way, let you meet your Ideal. I sat there and listened to his rudeness for a whole hour, and now you blame us!"

And she cried like a child. Everybody felt miserable, in a word, and the visit we had been looking forward to for so long and had prepared so far in advance left us very unhappy.

About five days later, however, Dostoevsky came to see us once again, and this time things worked out as beautifully as could be. Neither my mother nor my aunts were at home. My sister and I were alone together, and somehow the ice melted right away. Fyodor Mikhailovich took Anyuta by the hand, they sat down side by side on the couch and immediately began talking together like two old friends. The conversation didn't drag as it had on the previous occasion, moving stiffly from one boring subject to another. Now both Anyuta and Dostoevsky seemed in a rush to express their thoughts, interrupted each other, joked and laughed.

I sat there with them, not mixing into their discussion, not taking my eyes off Fyodor Mikhailovich and thirstily drinking in every word he said. This time he seemed an entirely different person, quite young and so simple, so nice and intelligent. "Is it possible that he's forty-three years old already?" I thought. "Is it possible that he's three and a half times as old as me and more than twice as old as Anyuta?[2] And on top of that, he's a great writer—and yet you can treat him like a friend!" And I felt immediately that he had become incredibly dear and close to me.

"What a lovely little sister you have!" said Dostoevsky suddenly and quite unexpectedly, although he had been speaking to Anyuta on quite a different topic only a minute before and seemed not to be paying attention to me at all.

I blushed all over with pleasure. My heart overflowed with gratitude to my sister when, in response, she told Fyodor Mikhailovich what a kind, bright girl I was, how I alone in the family sympathized with her and helped her. She grew quite animated in her praise of me and attributed to me unheard-of virtues. In conclusion, she informed

Dostoevsky that I wrote poetry: "Yes, really—and not bad at all for her age!"

And despite my weak protest she ran off and brought back a fat notebook filled with my doggerel, from which Dostoevsky, smiling slightly, immediately read two or three excerpts which he praised. My sister was aglow with pleasure. Lord, how I loved her at that moment! I felt that I would have given my life for these dear and precious people.

About three hours passed imperceptibly. Suddenly a bell sounded in the entrance hall. It was Mama, coming back from the Shopping Arcade. Not knowing that Dostoevsky was there with us, she came into the room still wearing her hat, all loaded down with her purchases and apologizing for being a little late for dinner.

Seeing Dostoevsky together with us there so unceremoniously, she was very much surprised and even frightened at first. "What would Vasily Vasilievich say to this!" was her first thought. But we threw our arms around her neck, and when she saw us so pleased and glowing she also melted, and it all ended with her inviting Fyodor Mikhailovich to stay and take pot luck with us.

From that day on he was very much at home in our house and, inasmuch as our stay in Petersburg would not last for long, he began coming to visit very often, three or four times a week.

It was especially pleasant if he came in the evening when there were no other outsiders except himself. At those times he would liven up and become extraordinarily dear and captivating. Dostoevsky hated general conversations. He spoke only in monologues, and then only on condition that all those present were people he liked, who listened to him with the closest attention. But then, if this condition was fulfilled, he spoke more beautifully, vividly and graphically than anyone else I have ever heard.[3]

Sometimes he would recount to us the plots of novels he

had been thinking about, sometimes scenes and episodes from his own life. I vividly recall, for instance, his description of the minutes when, as a condemned man sentenced to be shot, he had to stand up, his eyes already blindfolded, before a platoon of soldiers, waiting the fatal command to fire. And then suddenly a drum started to beat instead, and the news came that the sentence had been commuted.

I remember another story as well. My sister and I knew that Dostoevsky suffered from epilepsy, but in our eyes this disease was shrouded in such magical dread that we could never bring ourselves to touch on the subject even by the slightest hint.

To our amazement, however, he started to speak about it himself, and told us the circumstances of his first seizure. Later I heard another and quite different version of this episode, according to which he contracted epilepsy as a consequence of corporal punishment which he underwent as a convict in penal servitude. These two versions did not resemble one another at all. Which of them is the correct one I do not know, since many doctors have told me that almost all patients suffering from this disease present a typical feature: that they themselves forget how it began and constantly fantasize about it.[4]

Be that as it may, this is what Dostoevsky told us. He said that the disease had begun when he was no longer serving at hard labor but was already in Siberian exile. He was overwhelmed by the solitude at that time, and for months on end he did not see another living soul with whom he could exchange a sensible word. Suddenly and quite unexpectedly one of his old comrades came to see him (I forget now which name it was that Dostoevsky mentioned). It was just on the eve of Easter Sunday. But they forgot what night it was in their joy at seeing one another and spent it all at home talking, noticing neither the time nor their own fatigue, and growing drunk on their own talk.

They spoke about what was dearer to both of them than anything else—literature, art and philosophy; they touched, finally, on religion. The friend was an atheist, while Dostoevsky was a believer. Each of them was hotly convinced that he was right.

"There is a God—there is!" Dostoevsky finally shouted, beside himself with excitement. Just at that moment the bells of the neighboring church began to peal for holy matins. The very air resounded and vibrated.

"And I felt," said Fyodor Mikhailovich, "that heaven had descended to earth and swallowed me. I truly attained God and was penetrated by him. 'Yes, there is a God!' I cried out—and I remember nothing after that.

"All you healthy people," he went on, "don't even begin to understand what happiness is, the happiness that we epileptics experience during the instant before our attack. Mohammed maintains in his Koran that he saw Paradise and was in Paradise. All the clever fools are convinced that he was simply a liar and a fraud. Not at all! He was not lying. He actually was in Paradise in a fit of epilepsy, which he suffered with just as I do. I don't know whether that bliss lasts for seconds or hours or months, but believe me, I wouldn't exchange all the joys life can offer for that bliss!"

Dostoevsky said these last words in his characteristic passionate, spasmodic whisper. We sat there magnetized, spellbound by the fascination of what he was telling us. Suddenly an identical thought entered all our heads: he was going to have a fit right then and there.

His mouth twisted nervously, his whole face was twitching. He must have read the fear in our eyes. He abruptly stopped talking, passed his hand over his face and smiled sarcastically. "Don't be afraid. I always know ahead of time when it's coming."

We felt awkward and embarrassed that he had guessed our thoughts, and we didn't know what to say. He left soon afterwards and told us later that he really had suffered a severe seizure that very night.

There were times when Dostoevsky was very explicit in his speech, quite forgetting that he was in the presence of young ladies. My mother was sometimes appalled by the things he said. Once, for example, he began to narrate a scene from a novel he had conceived as a young man. The hero was a middle-aged landowner, highly cultivated, well-traveled, a reader of serious books and a purchaser of paintings and engravings. In his youth he had sowed his wild oats but then steadied down, acquired a wife and family, and enjoyed universal respect.

One morning he woke up early. The sun was peering into his bedroom window; the room was so neat, so pleasant and cozy. And he himself felt neat and estimable. A sense of satisfaction and repose flowed all through his body. Like a true sybarite he was in no hurry to wake up, he wanted to prolong this agreeable state of passive well-being.

Remaining at some midway point between sleep and wakefulness, he re-experienced in his mind various pleasant moments during his last trip abroad. He saw again the amazing strip of light falling on St. Cecilia's bare shoulders in the Munich museum. And there also entered his mind some very wise passages from a book he had read not long before, *On Universal Beauty and Harmony.*

All of a sudden, at the very zenith of these pleasant fantasies and memories, he began to feel a kind of discomfort—not quite inner pain, not quite anxiety. It was the sort of thing that occurs with people who have old, long-forgotten gunshot wounds from which the bullet has not been extracted: a moment ago nothing hurt and then suddenly the old wound starts to ache, and goes on aching, aching.

Our landowner began to consider and think things out. What could this mean? He was not in any pain, he had no sorrow of any kind. And yet he felt as though cats were scratching at his heart, and the feeling grew worse and worse.

It began to seem to him that there was something he had to call to mind, and now he made an effort to do this, he strained his memory. And suddenly he did remember, so vividly, so explicitly, and together with the memory there was such a sense of disgust through his whole being, that it was as though it had happened yesterday and not twenty years ago. And yet it had not troubled him in the slightest through all those twenty years. He remembered that once, after a night of debauchery, and egged on by his drunken companions, he had raped a ten-year-old girl.[5]

When Dostoevsky told us this, my mother could only throw up her hands helplessly. "Fyodor Mikhailovich! Have mercy on us! There are children here, after all!" she implored in a despairing voice.

I didn't even understand the sense of what he had said; I only guessed by my mother's indignation that it must be something dreadful. And yet Mama and Fyodor Mikhailovich soon became fast friends. My mother grew very fond of him even though there were times when she had to put up with a good deal from him.

Toward the end of our stay in Petersburg Mama planned to give a farewell party and to invite all our friends. Dostoevsky, of course, was also invited. For a long time he demurred, but Mama managed somehow to persuade him, to her later regret.

Our party turned out to be quite nonsensical. Since my parents had been living in the country for about ten years by then, they no longer had any real "circle" of their own. There were old acquaintances and friends whose lives had long ago diverged and gone their different ways. In the course of those ten years, some of these acquaintances had succeeded in making brilliant careers for themselves and climbing to a very high rung on the social ladder. Others, on the contrary, had slid down into impoverishment and privation, and dragged out their drab existences on the remote streets of Vasilevsky Island, barely managing to make ends meet. There were no shared interests among all these people, but almost everyone had accepted Mama's invitation and come to our party for old times' sake and for *"cette pauvre chere Lise."*

The company assembled at our house was a rather large and very motley one. Among the guests were the wife and daughters of a certain Minister. The Minister himself had promised to drop in for a few minutes toward the end of the evening, but failed to keep his word.[6] There was also a very old, bald and important official of German origin, of whom I remember only that he kept smacking the lips of his toothless mouth and kissing Mama's hand, saying every time, "She vas fery britty, your mother—much brittier as either of her daughters!"

There was a ruined landowner from the Ostsee province who now lived in Petersburg in the fruitless search for a remunerative position. There were many respectable widows and old maids and a few academicians, former friends of my grandfather's. Altogether, the predominating element was Germanic, proper, sedate, and colorless.

My aunts' apartment was very large, but it consisted of a multitude of tiny cell-like rooms overloaded with superfluous, unattractive knickknacks and trifles collected throughout the long lifetime of two elderly, tidy German maiden ladies. Because of the large number of guests and the many lighted candles, the air was suffocating. Two waiters in black tailcoats and white gloves handed round trays of tea, fruit and sweets. My mother, grown away from the life of the capital which she had once so dearly loved, felt an inner shyness and anxiety: was everything going off properly? Was it perhaps too out of date, too provincial? And might her former ladyfriends feel that she had fallen quite behind their social world?

The guests were completely uninterested in each other. They were bored, but as well-brought-up people for whom dull social evenings were an unavoidable ingredient of life, they submitted to their lot docilely and endured all this tedium with stoicism.

But one can imagine what happened to poor Dostoevsky when he fell into such a company. Both his appearance and his manner set him sharply apart from the rest. In an access of self-sacrifice he had thought it proper to dress in a tailcoat, and this tailcoat, which fit him both badly and clumsily, infuriated him inwardly throughout the evening. He had been in a bad temper from the moment he stepped across the threshold of our drawing room.

Like all highstrung people, he experienced an annoying shyness when he found himself in a group of strangers. The less he liked them, the more foolish and trivial they were, the greater was his unease. And he evidently needed to vent on someone else the irritation evoked by this feeling.

My mother was quick to introduce him to our guests, but instead of greeting them he muttered something inaudible that sounded like a grumble, and turned his back. Even worse, he immediately stated his claim to Anyuta's total attention. He took her off to a corner of the drawing room, revealing the obvious intention of keeping her there. This, of course, was contrary to all the social amenities. On top of that, his manner toward her was very far from what was acceptable in polite society. He took her hand. When he talked with her he bent down to her very ear.

Anyuta herself began to feel uncomfortable, and my mother was beside herself. At first she tried to convey to Dostoevsky "delicately" that his behavior was unacceptable. Passing by as though accidentally, she called my sister and tried to send her on some errand. Anyuta was on the point of getting up to go, but Dostoevsky very deliberately held her back, saying, "No, stay here, Anna Vasilievna, I haven't finished telling you my story."

At this, my mother lost her last shred of patience and flared up. "Excuse me, Fyodor Mikhailovich, but as a hostess Anyuta has the obligation to entertain our other guests too," she said very sharply, and took my sister away.

Dostoevsky was furious. Sulking in a corner, he maintained a stubborn silence and glared belligerently at everyone.

Among the guests was one who had made himself especially hateful to Fyodor Mikhailovich from the first moment. This was a distant relative of ours on the Shubert side—a young German and an officer of one of the Regiments of the Guard.[7] He was regarded as a very brilliant young man. He was handsome and intelligent and cultivated and received in the very highest society, and all this in the proper measure, in moderation and without excess. Even his career was proceeding in the proper measure, not with arrogant speed, but solidly, estimably. He knew how to make himself pleasing to the proper parties, but without over-eagerness or sycophancy.

By his rights as a relative, he paid a good deal of attention to his cousin Anyuta when he met her at the aunts', but this too was in proper measure, not conspicuously, but merely letting it be known that he had "intentions."

As always happens in such cases, everyone in the family knew that he was a potential and eligible suitor, but they all pretended not even to suspect such a possibility. Even my mother, when she was alone with my aunts, could not bring herself to touch upon this delicate matter except in hints and allusions.

Dostoevsky had only to take one look at this handsome, strapping, self-satisfied figure to conceive a dislike for him verging on frenzy. The young cuirassier, who had arranged himself picturesquely in an easy chair, was showing off in their full beauty his fashionably cut trousers, which tightly enclosed his long, well-shaped legs. Rustling his epaulets and bending lightly above my sister, he was telling her some very amusing story. Anyuta, still uncomfortable over the recent episode with Dostoevsky, was listening to him with her somewhat stereotyped drawing-room smile, "the smile of the gentle angel," as our English governess acidly used to call it.

Fyodor Mikhailovich looked at this pair, and a whole story composed itself in his mind: Anyuta loathes and despises this "cheeky little German," this "smug braggart," but her parents want to marry her off to him and are bringing them together in every way possible. The whole evening, of course, was arranged with this sole aim. Having invented his story, Dostoevsky believed it immediately and was filled with indignation.

The fashionable topic of conversation that winter was a book published by an English clergyman discussing the parallels between Russian Orthodoxy and Protestantism. In that Russo-German circle this was a theme of interest to all, and when the conversation touched on it the atmosphere livened up a little. Mama, herself of German origin, remarked that one of the advantages of the Protestants over the Orthodox consisted in the fact that they read the Gospel more.

"But was the Gospel written for society ladies?" suddenly blurted out Dostoevsky, who had remained stubbornly silent until then. "The Gospel says, 'First God created man and woman,' and further, 'Let a man forsake his father and mother and cleave to his wife.' That was how Christ understood the meaning of marriage! But what will the mamas say to that, when their only idea is how to marry their daughters off profitably?"

He uttered this with great intensity. As was habitual with him when he was upset, he seemed to shrink, and he shot out the words like bullets.

The effect was astounding. All the well-brought-up Germans fell silent and stared at him. Not until a few seconds had passed did they suddenly realize the awkwardness of what had been said, and then they all started talking at once, trying to drown it out. Again Dostoevsky glared at them belligerently and provocatively, and then he took himself back to his corner and didn't let out another word until the party was over.

The next time he appeared at our house Mama tried to receive him coldly, to show him that she was offended. But her uncommon kindness and gentleness never allowed her to stay angry at anyone for very long, least of all such a person as Fyodor Mikhailovich. Therefore they soon became friends again, and things resumed their old footing.

But Dostoevsky's relationship with Anyuta was somehow transformed after that evening, as though it had entered a new phase of its existence. He no longer overawed her in the least. On the contrary, she developed a desire to contradict him, to tease him. And he, for his part, began to show an extreme nervous tension and captiousness toward her, began demanding an account of how she spent her time when he wasn't there and behaving with hostility to anyone for whom she showed any admiration. He came to see us just as often as before, perhaps even more often, and he would stay longer than he used to do, but almost the whole time was spent in quarreling with my sister.

When they first met, Anyuta had been ready to forego any other pleasure, any invitation for the days when we expected a visit from Dostoevsky. If he was in the room she paid attention to no one else. But all this was changed now. If he came at a time when we had other visitors she would calmly go on entertaining her guests. If she was invited somewhere on an evening when it had been agreed that he would come to see her, she would write him a note making her excuses.

He would arrive the next day already resentful. Anyuta would pretend not to notice his bad humor. She would pick up a piece of needlework and begin to sew. This made Dostoevsky even angrier. He would sit down in a corner and remain sullenly silent. My sister would also be silent.

"Come on, drop that sewing!" he would say finally, not able to maintain his pose, and he would take the needlework out of her hands.

My sister would fold her hands submissively and say nothing.

"Where were you yesterday?" he would ask crossly.

"At a ball," she would answer with indifference.

"And did you dance?"

"Naturally."

"With that second cousin of yours?"

"Yes, with him and with others too."

"And does that amuse you?" the interrogation would continue.

Anyuta shrugged her shoulders. "For lack of anything better, even that is amusing," she would answer, and pick up her needlework again.

Dostoevsky looked at her for a few moments in silence. "Then you're an empty-headed girl, you're a foolish little brat, that's what you are!" he would conclude.

Their conversations frequently took on this tone now. A constant and very burning subject of the arguments between them was "Nihilism." Their discussions on this question sometimes went on long after midnight. The longer they talked the more excited they got, and in the heat of the argument they would express opinions much more extreme than either of them actually held.

"All the young people nowadays are stupid and uncultured!" Dostoevsky would shout. "For them, blacked boots are worth more than Pushkin!"

"Pushkin really is passé for our times," my sister would remark calmly, well aware that nothing in the world could infuriate Dostoevsky more than a disrespectful attitude toward Pushkin.

Beside himself with anger, Dostoevsky would sometimes pick up his hat and leave, stating solemnly that there was no point in arguing with a Nihilist, and that he wouldn't set foot in our house again. But the next day, of course, he would show up as though nothing had happened.

As the relationship between my sister and Dostoevsky was, to all appearances, deteriorating, my own friendship with him kept growing. With each passing day I admired him more and more and fell completely under his spell. He could not help noticing my boundless admiration, and he found it pleasant. He held me up as a constant example to my sister.

If he happened to express some profound idea or brilliant paradox which went counter to conventional morality, Anyuta would suddenly take it into her head to pretend not to understand. My eyes would blaze rapturously, but she, deliberately, in order to exasperate him, would offer some threadbare platitude in response.

"You have a worthless, petty little soul!" Fyodor Mikhailovich would flare up then. "Your sister is quite another thing! She is still a child, but how *she* understands me! Because she has a sensitive spirit!"

I would flush bright red with pleasure. I would have let myself be cut to pieces if necessary to prove to him how well I understood him.

Deep down, I was quite satisfied that Dostoevsky no longer showed my sister such admiration as he had when they first met. I was very much ashamed of this feeling and accused myself of a kind of betrayal of her. Making an unconscious bargain with my own conscience, I tried to atone for my secret sin against her through extra affection and helpfulness. But these gnawings of conscience did not prevent me from gloating against my will every time Anyuta and Dostoevsky quarreled.

He called me his friend. In utter naivete, I believed that I stood closer to him than my older sister did, and that I understood him better. He extolled even my looks, to Anyuta's detriment.

"You imagine that you're beautiful," he would tell her. But that little sister of yours will be much better looking with time. Her face is more expressive—she has gypsy eyes! And you—you're no more than a pretty little candy-box German girl, that's all!"

Anyuta would give him a disdainful smile. But I would rapturously drink in all these unprecedented praises of my beauty.

"And maybe it's even true!" I told myself with a sinking of the heart. And I began to feel quite seriously disturbed by the notion that my sister's feelings might be hurt by the preference Dostoevsky showed me.

I was dying to know for sure what Anyuta herself thought of all this, and whether it was really true that I would be a pretty girl when I was quite grown up. I was particularly absorbed by this latter question.

My sister and I shared a bedroom while we were staying in Petersburg. Our most heart-to-heart talks took place in the evenings when we were getting ready for bed. As was her habit, Anyuta stood before the mirror combing her long fair hair and plaiting it into two braids for the night. This was time-consuming work. Her hair was very thick and silky, and she passed her comb through it lovingly.

I was sitting on the bed undressed and ready for the night, with my arms hugging my knees, deciding the best way to open this conversation, so interesting to me.

"What funny things Fyodor Mikhailovich was saying today!" I began at last trying to appear as casual as possible.

"What things?" absentmindedly asked Anyuta who, incidently, had quite forgotten the conversation I found so important.

"About my eyes being gypsy eyes, and that I'm going to be pretty," I said—and felt myself blushing to my ears.

Anyuta lowered the hand holding the comb and turned her face toward me, gracefully bending her neck.

"So you believe that Fyodor Mikhailovich thinks you're pretty—prettier than me?" she asked, giving me a sly, enigmatic look.

That treacherous smile, those laughing green eyes and long blonde tresses made her look exactly like a water nymph. Next to her, I saw my own short, dark-skinned figure in the big pierglass standing opposite her bed, and I could compare the two of us. I cannot say that I found this comparison very pleasant, but my sister's cold and self-assured tone of voice made me resentful, and I didn't want to give in.

"There are different kinds of tastes!" I said angrily.

"Yes, there are odd tastes," Anyuta calmly answered, and went on combing her hair.

After the candle was snuffed out I lay with my face buried in the pillow, and still I went on with my reflections on the same theme. "But perhaps Fyodor Mikhailovich does have the kind of taste that likes me better than my sister," I thought to myself. And then, from the mechanical habit of my childhood, I started to pray inside my head: Dear God, oh God: please, please, let the whole world admire Anyuta, but just fix things up so that Fyodor Mikhailovich will think I'm prettier! My illusions on this point, however, were destined to be harshly shattered in the very near future.

Among those agreeable talents whose cultivation Dostoevsky encouraged was music. Up to then I had taken piano lessons the same as most girls do, without any special partiality or dislike for them. My ear for music was mediocre, but inasmuch as I had been made to practice my scales and exercises for an hour and a half every day since I was five years old, by the age of thirteen I had managed to develop a certain finger dexterity, a passable touch and the ability to sight-read quickly.

Once, at the very beginning of our acquaintance, I had played for Dostoevsky a piece that I always carried off very well: a set of variations on Russian folk themes. Fyodor Mikhailovich was no musician. He belonged to that category of people for whom pleasure in music depends on purely subjective causes, on the mood of the moment. At times the most splendid and beautifully performed music will evoke no more than a yawn from such people; while at other times a barrel organ squalling away on the street will move them to tears.

On the day when I first played the piano for him, Fyodor Mikhailovich happened to be in just such a receptive, easily moved frame of mind, and therefore he waxed ecstatic over my playing. Carried away, as was his tendency, he lavished on me the most exaggerated praises: my talent, my "soul," and goodness only knows what else.

From that day on, of course, I developed a passion for music. I wheedled my mother into getting me a good piano teacher and I spent every free minute at the piano throughout our stay in Petersburg, so that during those three months I really did make substantial progress.

Now I prepared a surprise for Dostoevsky. He had once told us that of all musical works, his favorite was Beethoven's *Sonata Pathétique,* and that this sonata never failed to plunge him into a whole world of forgotten sensations. Although the sonata was much more advanced than anything I had played up to that time, I resolved to learn it no matter what. And, as it turned out, after expending untold hours of labor on it, I reached the point of playing it fairly tolerably. Now I awaited only a fitting occasion to gladden Dostoevsky with it. That occasion presented itself very soon.

Only five or six days were left before our departure for home. Mama and all the aunts had been invited to an important dinner party at the Swedish embassy, for the Ambassador was an old friend of the family. Anyuta, who by this time had had enough of visits and dinner parties, excused herself on the ground of a headache. The two of us were alone in the house. That evening Dostoevsky came to see us.

The imminence of our departure, the awareness that none of our elders was at home and that an evening like this would not come again soon—all this put us into a mood of pleasant excitement. Fyodor Mikhailovich, somehow, also seemed different: nervous but not irascible as he had so often been of late—on the contrary, gentle and affectionate.

So this, then, was the perfect time to play him his favorite sonata! I rejoiced in advance at the thought of all the pleasure I was going to give him.

I began to play. The difficulty of the piece, the necessity of following every note, the fear of striking a wrong note soon engulfed my attention so completely that I blotted out my surroundings and didn't notice anything that was happening around me. I finished the sonata with the self-satisfied awareness of having played well. There was an enjoyable tiredness in my fingers. Still under the spell of the music and the stimulus of the pleasurable excitement which always takes hold of one after a piece of work well done, I waited for my well-deserved praise. But there was only silence. I looked around: there was no one in the room.

My heart sank. Having as yet no definite suspicion but feeling a dim presentiment of something wrong, I went into the next room. It, too, was empty. Finally I lifted the portiere draped over the door to a little corner salon, and saw that Fyodor Mikhailovich was there with Anyuta.

But Lord, what did I see!

They were sitting side by side on a small settee. The room was dimly lit by a lamp with a big shade. The shadow fell directly on my sister so that I couldn't make out her face, but I saw Dostoevsky's face distinctly. It was white and agitated. He was holding Anyuta's hand in his. Leaning toward her, he spoke in the same passionate, spasmodic whisper I knew and loved so well.

"My darling Anna Vasilievna, try to understand . . . I fell in love from the first minute I saw you. Even before that. I had an intimation even from your letters. And it's not in friendship that I love you but in passion, with all my being."

My eyes blurred. A feeling of bitter loneliness, of deadly insult suddenly gripped me and there was a rush of blood, first, it seemed, to my heart, and then surging in a hot stream to my head.

I dropped the curtain and ran out of the room. I heard a noise as a chair I had accidentally bumped into fell over.

"Is that you, Sonya?" my sister called out in alarm. But I didn't answer her and didn't stop running until I reached our bedroom at the far end of the apartment, at the end of a long corridor. Once there, I immediately and hurriedly started to undress. Without lighting the candle I tore off my dress, and still half-clothed I threw myself into bed and buried my head under the blanket. I had only one fear at that moment: that my sister might come after me and call me back into the drawing room. I couldn't bear to look at them now.

Feelings I had never experienced before filled and overflowed my heart: bitterness, hurt, shame. Mostly, it was shame and hurt. Up to that moment I had not acknowledged even in my most secret thoughts how I felt about Dostoevsky and had not admitted even to myself that I was in love with him.

Although I was only thirteen years old I had already read and heard a lot about love. But it seemed to me somehow that people fell in love in books and not in real life. As far as Dostoevsky was concerned, I imagined that things would go on this way forever, all our lives, just as they had been during these past months.[8]

And suddenly, all at once, it's finished—all finished! I kept thinking to myself in despair. And only now, when everything seemed irretrievably lost, did I realize how happy I had been during all these days: yesterday, today, a few minutes ago. And now—Lord—now!

Precisely what was finished, what was changed, I couldn't tell myself directly even then. I only felt that everything was withered for me, that life wasn't worth living.

And why did they make a fool out of me? I reproached them with unjust anger. Why were they so secretive about it? Why did they pretend? Well then, let him love her, let him marry her, I told myself a few seconds later—what's it to me? But the tears kept flowing, and deep inside I felt the same new, unbearable pain.

Time passed. Now I would have liked Anyuta to come after me. I was indignant with her for not coming. I might as well be dead for all they cared! God, God! If I could really die now! And all of a sudden I felt inexpressibly sorry for myself, and the tears flowed harder.

What were they doing now? How lovely it must be for them! And at this thought I got a frenzied desire to run straight to them and say a lot of insolent things.

I jumped out of bed, and with hands trembling in excitement I began groping for the matches to light the candle, so that I could start dressing again. But the matches were not to be found. Since I had scattered my clothing all over the room I couldn't dress in the dark, and I was too embarrassed to call the maid. Therefore I fell back into bed and started sobbing all over again, with a feeling of helpless, hopeless loneliness.

When the organism is unused to suffering, the first tears exhaust one very quickly. My paroxysm of sharp despair changed into a dull numbness. Not a sound carried from the front rooms to our bedroom, but I could hear the servants getting their supper ready in the kitchen nearby. There was a clatter of knives and plates; the maids were laughing and talking. Everybody was gay, everybody was feeling good . . . everybody but me.

At last, after what seemed like several eternities, there was a loud ring of the doorbell. It must be Mama and the aunts, back from their dinner party. I could hear the quick steps of the footman going to open the door; then loud, cheerful voices sounded in the entrance hall, as always when people come back from a visit.

Dostoevsky was probably still there. Would Anyuta tell Mama today, or would she wait till tomorrow? And then I could distinguish his voice from among the others. He was saying goodnight, hurrying to leave.

Straining to hear, I could even make out the sound of his putting on his overshoes. Then the front door banged again, and soon after that Anyuta's steps rang down the corridor. She opened our bedroom door, and a brilliant strip of light fell directly on my face. This light was unbearably bright, offensive to my eyes swollen with crying, and a feeling of physical aversion for my sister rose up to my throat.

Horrid girl! She's gloating, I thought bitterly. I quickly turned my face to the wall and pretended to be asleep.

Anyuta unhurriedly set the candleholder down on the chest of drawers, then walked up to my bed and stood there for a few moments in silence.

I lay without stirring, holding my breath.

"But I can see you're not asleep!" she said finally.

I didn't answer.

"All right then, if you want to sulk . . . sulk! So much the worse for you. You won't find out anything!" And she began undressing as though nothing had happened.

I remember that I dreamed a marvelous dream that night. It's a strange thing in general—whenever in my life a deep, overwhelming grief descends upon me, afterwards, that night, I always have extraordinarily happy and pleasant dreams.

But then how painful is the moment of awakening! The dreams have not yet quite evaporated; all through my body, weary from yesterday's tears, I feel a pleasant exhaustion after a few hours of pleasurable sleep, the physical satisfaction of harmony restored. Then suddenly the memory of the horrible, irreparable thing that happened the day before pounds in my head like a hammer, and my heart is gripped by the knowledge that it is necessary to begin living and suffering again.

There is much in life that is distressing. Every form of suffering is repugnant. The paroxysm of the first wild despair is painful, when all one's being is outraged and does not want to submit and cannot apprehend the full weight of one's loss. But hardly better are the long, long days which follow, when the tears are all cried out, and the outrage is subdued, and the person stops beating his head against the wall but is aware only of the slow process of disintegration and decay, invisible to others, taking place in his heart under the weight of the grief which has befallen him. All this is very painful and agonizing. But nonetheless, the first moments of return to melancholy reality after a short interval of unconsciousness—that is perhaps the most painful thing of all.

I spent the whole next day in feverish expectation. What would happen now? I asked my sister no questions. I continued to feel the same dislike of her that I had felt the day before, perhaps a degree less sharp, and therefore I avoided her as much as possible.

Seeing me so miserable, she wanted to come up to me and caress me, but in the wrath that suddenly took hold of me I roughly pushed her away. Then her feelings, too, were hurt, and she left me to my own disconsolate thoughts.

For some reason I expected that Dostoevsky would

surely come to us today and that then something terrible would happen, but he did not come. We were already sitting down to dinner, and he had not yet appeared. And I knew we were supposed to go to a concert that evening.

As time went on and he did not come, I started feeling a little better. I even began to feel a kind of vague, indeterminate hope. Suddenly it occurred to me that my sister would back out of the concert and stay at home, and that Fyodor Mikhailovich would come to her when she was alone. My heart pinched with jealousy at this thought.

But Anyuta did not back out of the concert. She came along with us and was very merry and talkative all evening. When the concert was over and we were at home getting ready for bed, and Anyuta was about to blow the candle out, I couldn't hold back any longer. Without looking at her I asked, "But when is Fyodor Mikhailovich coming to see you?"

Anyuta smiled. "But you don't want to know, you don't want to talk to me, you'd rather sulk."

Her voice was so soft and kind that my heart melted at once, and again she became very dear to me. How could he not love her when she's so wonderful, and I'm so nasty and mean? I thought, in a surge of self-abasement.

I crawled into bed with her, nestled up to her and started to cry. She stroked my hair.

"All right now, little silly, stop now! What a silly little goose you are!" she repeated affectionately. But then she couldn't restrain herself and burst into uncontrollable laughter.

"She went and fell in love, and with whom? With a man three and a half times her age!"

These words, this laughter suddenly awakened in me a wild, all-encompassing hope.

"Then you don't really love him?" I whispered, almost suffocating with excitement.

Anyuta grew thoughtful. "Well, it's like this . . ." she

began, evidently searching for the right words and having difficulty finding them. "Of course I love him very much and I respect him, I respect him terrifically! He's so kind, so intelligent . . . he's a genius!"

She was quite animated, and again I felt my heart pinch. "But . . . how shall I explain it to you? I don't love him the way that he . . . well, anyway, I don't love him the way you love somebody you want to marry!" she suddenly concluded.

God! How my heart filled with light! I flung myself at Anyuta and started kissing her hands and neck. She went on talking for a long time.

"You see, I'm even surprised myself sometimes that I can't love him! He's such a wonderful person. In the beginning I thought I might come to love him. But he needs an entirely different kind of wife from me. His wife will have to dedicate herself to him utterly, utterly, to give up her whole life to him, to think about nothing but him. And I can't do that . . . I want to live myself! And then, he's so nervous and demanding. He always seems to be taking possession of me and sucking me up into himself. When I'm with him I can never be myself."[9]

Anyuta said all this as though addressing it to me, but in reality she was explaining it to herself. I put on an air of understanding and sympathy, but in my heart I thought, "God! What happiness it would be to be with him constantly, and submit yourself to him utterly! How could my sister push such happiness away?"

Be that as it may, I fell asleep that night not nearly so wretched as the day before.

And now the day of our departure was quite close. Fyodor Mikhailovich came to see us once more to say goodbye. He didn't stay very long, but he behaved in a friendly, simple way with Anyuta, and they promised to write one another. His goodbye to me was very tender. He even kissed me at parting but was probably very far from suspecting the nature of my feelings toward him and the depth of the suffering he had caused me.

About six months later[10] my sister received a letter from him in which he told her that he had met a remarkable young woman whom he loved and who had consented to marry him. This young woman was Anna Grigorievna, his second wife. "If anyone had foretold this to me half a year ago," he naively remarked at the end of the letter, "I swear on my honor that I wouldn't have believed it."

My own wounds also healed quickly. During the few days left in Petersburg I still felt an unprecedented weight on my spirit and went about more dejected and subdued than usual. But the trip home wiped clean the last traces of the storm I had just experienced.

We left in April. It was still winter in Petersburg, cold and nasty. But in Vitebsk we were met quite unexpectedly by real spring, which in the course of some two days had taken claim to all its rights. All the streams and rivers flooded their banks and overflowed like seas. The frozen earth was thawing. The mud was indescribable.

We managed somehow, as long as we were driving along the highway. But when we reached our provincial town, we were compelled to leave our traveling coach at an inn and to hire two shabby old springless carriages. Mama and the coachman drove and worried . . . how in the world would we make it home? Mama worried mostly about my father's scolding her for overstaying our time in Petersburg. But in spite of all the groaning and complaining, the drive was marvelous.

I remember crossing through a dense pine forest late one evening. Neither my sister nor I could sleep. We sat there not speaking, reliving once more all the different experiences of the past three months and greedily drawing into our lungs the spicy smell of spring that permeated the air. Our hearts ached painfully, with a kind of langourous expectation.

Little by little, it grew quite dark. Because the road was so bad, we drove at a very slow pace. The coachman seemed to have dozed off on his coachbox and was not urging the horses on. All that could be heard was the clip-clop of their hooves through the mud and the faint, irregular tinkling of the carriage bells. The forest stretched out on both sides of the road, dark, mysterious, impenetrable.

Suddenly, as we came out into a clearing, the moon seemed to float out from behind the woods and flooded us with silvery light, so brilliantly and unexpectedly that it gave us an eerie feeling. After the last talk with my sister in Petersburg we no longer touched on intimate matters. There was still some kind of constraint between us, some new thing was dividing us.

But now, at that moment, as if by unspoken agreement, we clung together, we put our arms around each other. And both of us felt that there was no more estrangement between us, that we were as close as we had been before. A feeling of inexplicable, boundless joy took hold of us both. God! How that life stretching out before us drew us to it, and beckoned us on, and how boundless, mysterious and beautiful it appeared to us on that night!

Notes

Chapter One: *Earliest Memories*

1. The family name of Sofya's father is encountered in three variant forms: Kryukovskóy (with the accent on the final syllable), Krukóvsky (accented on the middle syllable), and Korvin-Krukovsky. The first form was the one used by the inhabitants of the region where the family estate, Palibino, was located, near the Lithuanian border in the then province of Vitebsk. It is still in use today, although the name Palibino is now known as Polibino. The second form, Krukovsky, was the one in general use by the family up to the year 1858; this is the name used in recording Sofya's birth and baptism at the Znamenskaya Church in Moscow on January 15, 1850.

The Krukovsky family made repeated applications to the Department of Heraldry for confirmation of the family's right to be considered as belonging to the ancient nobility. These petitions were refused until after General V. V. Krukovsky's retirement from the Army in 1858 with the rank of Lieutenant-General of Artillery, at which time the petition was finally granted. The decree granting the petition uses the name Korvin-Krukovsky. The name of Korvin is associated with Matthias Corvinus, King of Hungary. It appears on the genealogical tree which adorned the family library at Palibino, according to which a daughter of Matthias Corvinus was married to a certain Polish hero, Krukovsky. But no actual evidence authenticating this version of the family's lineage has been found in official archival records.

2. General Krukovsky served in Kaluga from 1855 to 1858.

3. Anyuta (Anna Vasilievna Korvin-Krukovskaya) was born in 1843 and died in 1887.

4. Fedya (Fyodor Vasilievich Korvin-Krukovsky), five years younger than Sofya, was botn in 1855 and died in 1919. After completing his studies at the Physics-Mathematics Faculty of the University of Petersburg in 1878, he served in one of the government ministries. He married late in life and had one child, a daughter. He never evidenced any of the creative vitality of his two sisters. The family tutor Malevich has some interesting things to say about Fedya's upbringing as it influenced his character development (see note 5 to Kovalevskaya's *An Autobiographical Sketch,* 229).

Chapter Three: *Metamorphosis*

1. Kovalevskaya's manuscript reads "a Polish tutor" without mentioning him by name. Yosif Ignatievich Malevich (1813–1898) lived at Palibino for nine years. His own reminiscences take sharp issue with

Kovalevskaya's assessment of his minimal influence on her education. Indeed, Kovalevskaya herself wrote a reminiscence entitled *From the Time of the Polish Uprising* which clearly shows how Malevich's teaching molded her sympathy for the cause of the Polish insurgents against their Russian conquerors. (The article was published in Sweden in *Nordisk tidskrift* under the pseudonym of Tanja Rajevski. It was, of course, unpublishable in Tsarist Russia.)

2. Miss Margaret Smith (1826–1914) belonged to an English family which had lived in Russia for many years. Her charges addressed her in the Russian fashion as "Margarita Frantsevna" or (behind her back) as "Margarita." After leaving the Korvin-Krukovskys she went to live with the Shubert family in Petersburg.

CHAPTER FOUR: *Palibino*

1. This chapter is published here for the first time in English translation from Kovalevskaya's original Russian text. It did not appear at all in Russian until the edition of 1951, edited by S.Ya.Shtraikh. At that time the chapter was translated from the Swedish by Kovalevskaya's daughter Sofya Vladimirovna. Later, Kovalevskaya's original Russian manuscript was discovered in the archives of the USSR Academy of Sciences. The text included many variant readings and inserts, some of them written in French in Kovalevskaya's hand and evidently intended for a French edition of the work.
The chapter as given here is translated from Kovalevskaya's text except for the last seven lines, beginning "But the concert of the wolves had a terrifying effect upon the dog. . . ." which remain in Sofya Vladimirovna's translation.

2. As has been mentioned, the Swedish edition of the work appeared not as a memoir but in the form of a novel narrated in the third person; the family name was changed from Korvin-Krukovsky to Rajevski. In Soviet editions the family name is changed back to Korvin-Krukovsky and Tanja is again Sofya (or Sonya, the affectionate form), but the third person is retained. The present edition carries the change one logical step further: it switches to the first person so as not to disrupt the mood and tempo of the rest of the narrative. The title "Palibino" was supplied by S.Ya.Shtraikh.

3. A schismatic sect which arose in the latter part of the 17th century in opposition to reforms in Russian Orthodox ritual carried out by the Patriarch Nikon.

Chapter Five: *Miss Smith*

1. The reference is to *A Russian Anthology* (with Commentaries, Compiled by Andrey Filonov. Vol. 1, "Epic Poetry," Vol. 2, "Lyric Poetry," Vol. 3, "Dramatic Poetry." St. Petersbury, 1863). By "poetry" the compiler understood imaginative literature in general. Malevich also mentions this book in his reminiscences. However, the book contains neither Lermontov's *The Novice* nor Pushkin's *Captive of the Caucasus.* It is possible that Kovalevskaya had access to another widely disseminated textbook of that period: *A Complete Russian Anthology,* compiled by A. Galakhov, Moscow, 1857.

2. These childish verses have been lost. Very little of Kovalevskaya's verse has been located up to the present time.

3. *Undine* (1811), a long narrative fairy tale by the German writer Friedrich de la Motte Fouqué, freely translated into Russian by V. A. Zhukovsky. *The Novice* (Mtsyry, a Georgian word) by Mikhail Lermontov, a long narrative poem closely related in meter and diction to Zhukovsky's translation of Byron's *Prisoner of Chillon.*

4. This is not quite accurate. As S.Ya. Shtraikh points out, Elizaveta Fyodorovna's diaries dating from the first year of her marriage up to 1851 are filled with bitter complaints against Vasily Vasilievich's behavior toward her.

Chapter Six: *My Uncle Pyotr Vasilievich Krukovsky*

1. In actuality he had two sons, Andrey and Alexander. Kovalevskaya's description of her uncle contains several errors.

2. Paul Bert (1833–86), French physiologist and zoologist, student and collaborator of Claude Bernard, worked in animal transplantation not for the purpose of experimental surgery, but as a contribution to the physiological problem of the adaptation of transplanted organs and tissues to a new environment. The experiment referred to here is the so-called "double monster," in which two rats were united by suturing their skins together.

3. The reference is to Helmholtz's researches on the conservation of energy.

4. Actually he was an artillery officer for a time, a second lieutenant in the First Army. He retired from army service in 1826.

5. Mikhail Vasilievich Ostrogradsky (1801–61), member of the Petersburg Academy of Sciences and author of numerous articles on mathematical analysis and its application.

6. Kovalevskaya was then about eleven years old.

7. There is a discrepancy here. In her *Autobiographical Sketch* (see p. 213) Kovalevskaya says that she studied calculus with Strannolyubsky the year before her marriage; that would make her seventeen and not fifteen.

8. Alexander Nikolayevich Strannolyubsky (1839–1903) was a well-known teacher of mathematics and a strong proponent of higher education for women (he taught for the Alarchin Women's Courses). An educational innovator, he opposed the use of compulsion and reward in educating the young, encouraged the teaching of craft skills, introduced field trips to factories into the school curriculum. He was extremely popular in Petersburg radical youth circles during the 1860s and 70s.

CHAPTER SEVEN: *Uncle Fyodor Fyodorovich Shubert*

1. Born in 1831: hence, he was about 28 years old at the time of the episode described. He completed his university education with the degree of Kandidat (roughly equivalent to an American doctorate) and then served in one of the offices of the War Ministry. He died childless in 1877.

2. The preceding three paragraphs, beginning "Preserved gooseberries . . ." and ending "And this mollified me a little" have been inserted from the Swedish edition, with third person changed to first and Tanja to Sonya.

CHAPTER EIGHT: *My Sister*

1. The Polish uprising of 1863 was the third since Poland's three partitions by Russia, Austria and Prussia (in 1775, 1793 and 1797), and the one most violently suppressed by the Russian rulers. The first abortive revolt, led by the American Revolutionary War hero Thaddeus Kosciuszko, lasted perhaps two weeks. In 1815, during the reign of Alexander I, the Kingdom of Poland—while geographically a part of the Russian Empire—still retained a degree of autonomy in its internal life and possessed a liberal constitution. The Polish insurrection of 1830, an effort to win back Poland's lost provinces and its independence, was sufficient excuse for Nicholas I, Alexander's successor, to abolish the constitution and place Poland under a military regime for the remainder of his reign. After the uprising of 1863, ferociously suppressed by Muravyov ("the Hangman"), the imperial policy of compulsory Russification was applied with particular heavy-handedness

to Poland, which was placed under enormous pressure to abandon its national traditions and language and to acknowledge the superiority of Russian culture.

2. Elizaveta Fyodorovna's diary bears witness to Anyuta's emotional state during this period:

> Anyuta is bored, is wishing for she doesn't know what, those "delights of life" she has not as yet experienced. Even though I don't approve of her views on life, when I look at her I understand the dreams and urges of youth, which cost me so dearly. (June 23, 1864)

3. Kovalevskaya's account differs considerably from the denouement as given in Bulwer-Lytton's novel.

CHAPTER NINE: *Anyuta's Nihilism*

1. The Znamenskaya Commune (so-called because it was located on Znamenskaya Street) did actually exist for a short time and was run by a group of young men and women who pooled their modest resources and tried to care for themselves without the help of servants. The commune was founded by the radical writer Vasily Sleptsov (1836–78), author of a number of novels and sketches. Sleptsov's activity was at its height between 1861 and 1866, at which time he was arrested and imprisoned. He was forced to give up literary work in the 1870s because of poor health.

2. These "great men" are named in the Swedish edition: Chernyshevsky, Dobrolyubov, Sleptsov.

3. *The Epoch* (Epokha) was founded by Fyodor Dostoevsky together with his brother Mikhail in 1863 as the successor to their ill-starred *Time* (Vremya), which had been closed down by imperial decree at the time of the Polish uprising because of an article in it by Nikolai Strakhov, misterpreted by the authorities as constituting a glorification of Polish culture at the expense of Russian. *The Epoch* was extremely short-lived. In 1864 (the year when Anyuta's two short stories were published in it) Mikhail Dostoevsky died. The next year saw the journal's demise. Internal evidence, therefore, suggests that Anyuta's first serious involvement with the "Nihilist" ideas described in this chapter began in 1862 or 1863, when she was about twenty and Sofya thirteen years old.

4. *The Contemporary* (Sovremennik) was founded in 1836 by Alexander Pushkin and had an ideologically varied and relatively long history. Under the influence of Chernyshevsky (who joined the staff in 1854) and Dobrolyubov (1856) it became the rallying-ground of the

extreme left. It was closed down by Tsarist decree in 1866.

The Russian Word (Russkoye slovo) was the most outspokenly revolutionary of the journals mentioned here. It too was shut down in 1866 because of its "corrupting influence on youth."

The Bell (Kolokol) was published in London (1857–65) by the exiled Alexander Herzen and Nikolai Ogarev and secretly smuggled into Russia, where it exercised enormous influence and had a broad spectrum of readers up to and including the Tsar. It was moved to Geneva in 1865 as part of Herzen's "Free Russian Press"; it lasted until 1867.

CHAPTER TEN: *Anyuta's First Literary Experiments*

1. Kovalevskaya's memory betrays her here. Although she conveys the gist of Dostoevsky's letter to Anyuta, she does not reproduce it verbatim. It should be remembered that there was an interval of twenty-two years between Dostoevsky's letter and Kovalevskaya's reminiscence.

2. It appeared in *The Epoch,* No. 8, for August, 1864.

3. To understand the remarkable fervor with which Dostoevsky responded to Anyuta's unremarkable story, one should recall the dismal circumstances of his life at the time he received it. His wife, Maria Dmitrievna Isayeva, had died of tuberculosis three months before, in April, leaving a son Paul by her first husband and binding Dostoevsky with a promise to look after the youth, who was then eighteen years old. The affair with Polina Suslova was dragging to a conclusion but was not quite finished. *The Epoch* was on the verge of financial collapse. Mikhail, its joint owner, died suddenly in July, bequeathing to his brother a mountain of debts and a large and helpless family. For all of these burdens Dostoevsky assumed full financial and moral responsibility. That he should have reacted with deep depression is not surprising. This was the situation when Anyuta's youthful story and letter arrived. Dostoevsky's interest was immediately aroused, perhaps less by the story than by its author. Not only did he publish her story in the next issue of *The Epoch,* he also paid her honorarium at once, although the journal's financial plight was such that other better-known and more regular contributors went unpaid until long afterward. Dostoevsky maintained a secret correspondence with Anyuta not only through the Krukovskys' housekeeper, as Kovalevskaya relates, but also through Anyuta's friend Zhanna Evreinova, another budding Nihilist who happened to be the daughter of the Imperial Palace commandant, General A. M. Evreinov.

4. Kovalevskaya's manuscript contains the following passage which did not appear in print:

> For my poor father, a female writer was the embodiment of every possible abomination. His attitude toward them was one of naive horror and indignation; he considered them, every one of them, capable of all kinds of wrongdoing. This attitude of his came to mind spontanously when I read Nekrasov's description of one of his characters: "He harshly criticized George Sand/For wearing trousers . . ."

5. The poetess Evdokia Petrovna Rostopchina (1811–59). She was commended by Pushkin, Zhukovsky and Lermontov for the elegance and sonority of her verse. In 1845 she published a controversial poem, "The Forced Marriage," in which the Russian state was depicted as a despotic husband who brutalized Poland, his oppressed bride-victim.

6. From a poen by N. A. Dobrolyubov which appeared in *The Contemporary* in 1862: "Then let me die, there's little grieving . . ." The lines quoted by Kovalevskaya are not quite verbatim.

7. This paragraph, taken from Kovalevskaya's draft manuscript, did not appear in print.

8. It was published in *The Epoch,* No. 9, September, 1864. The story's original title, "The Novice," as Dostoevsky wrote Anyuta on December 14, 1864,

> . . . was not exactly prohibited but was rejected by the ecclesiastical censor. At first he prohibited the story itself, and therefore I was compelled to assent to many deletions and revisions. Some of these revisions were necessary in my own judgment as well. . . . Let me add that the highest ability of a writer is the ability to cross out. He who has the ability and the strength to strike out his *own* writing will go far. All the great writers wrote with extreme compression. And the main thing is not to repeat what you have already said and what was clear without repeating it. . . . Our regular editorial staff and all those close to it liked your story very much. Not all of them liked "The Dream." My own opinion you already know. You not only may, you *must* take your talent seriously. You are a poet. That alone is worth a great deal. . . . One thing: study and read. Read serious books. Life will do the rest.

The story appeared in first place, as the opening item of the issue. Such an honor was a rare occurrence even for established writers.

9. An unpublished paragraph from Kovalevskaya's manuscript reads:

> "Sometimes Elizaveta Fyodorovna came to her daughter and tried to bring her round. "Anyutochka, come on, give Papa a little pleasure. Promise him not to write any more, and occupy yourself with something else. Look here, I remember that when I was a

young girl I suddenly got the desire to study the violin. But my father wouldn't let me do it. He considered it very ungraceful for a girl to handle a bow. Well—so what! Naturally I didn't insist, and I started taking singing lessons instead. So then why can't you give up this horrid literature and take up something else?"

10. A letter from General Korvin-Krukovsky to Dostoevsky dated January 14, 1866 (about a year after the first meeting between Dostoevsky and the Korvin-Krukovsky sisters) makes it clear that the General never did make the writer's acquaintance and never veered from his original assessment of him as "a journalist and ex-convict."

11. Kovalevskaya's draft manuscript adds a final line which she did not publish: "But I am afraid to remember it. Once you start thinking back, you cannot go on."

CHAPTER ELEVEN: *Our Friendship with Fyodor Mikhailovich Dostoevsky*

1. February 28, 1865.

2. Sofya was then fifteen years old, Anyuta twenty-three.

3. Kovalevskaya's draft manuscript contains a lengthy digression at this point in which she retells the highlights of Dostoevsky's early life: his success with his first novel *Poor Folk,* his sponsorship by the critic and literary arbiter V. V. Belinsky, his analysis of the Russian literary scene during that period, his involvement with the Petrashevsky circle, his arrest and condemnation to death. The section was evidently intended for a foreign readership; it appeared in the Swedish edition in abbreviated form.

4. A great deal has been written about the origins of Dostoevsky's epilepsy, including Sigmund Freud's essay, "Dostoevsky and Parricide," which attempts to establish a connection between the onset of the disease and the murder of Dostoevsky's father by his own serfs. The most recent and impassioned counter-argument to this thesis is given in Joseph Frank's *Dostoevsky: The Seeds of Revolt 1821–1849* (Princeton, 1976). Orest Miller, Dostoevsky's first biographer (St. Petersburg, 1883) says that the disease, contrary to Dostoevsky's assertion, manifested itself even before his arrest and imprisonment, but was not acknowledged by Dostoevsky until it erupted so violently that there could no longer be room for doubt as to the diagnosis.

5. This theme was used by Dostoevsky in *The Possessed* (in the chapter titled "At Tikhon's" in Russian, "Stavrogin's Confession" in English). The chapter was not published in Dostoevsky's lifetime. Nikolai Stra-

khov, Dostoevsky's putative "best friend," spread a scurrilous rumor in a letter to Tolstoy written years after Dostoevsky's death, alleging that the perpetrator of the rape was Dostoevsky himself (for Strakhov's letter and Anna Dostoevskaya's reply to it, see *Dostoevsky: Reminiscences,* by Anna Dostoevsky, trans. and ed. by Beatrice Stillman, Liveright-Norton, New York, 1975, pp. 371–382). The curious history of the suppressed chapter is given in the same volume, p. 412.

6. D. A. Milyutin, Minister of War during the 1860s and 1870s. Kovalevskaya's parents and later she herself maintained a warm relationship with the Milyutin family.

7. Kovalevskaya's manuscript contains the following notation: "Distant relative: a colonel of the General Staff, Andrey Ivanovich Kosich. Not for publication." In the published text all the passages in which Kosich is mentioned were changed, and in print he was transformed into "a German." Kosich was Chief of Staff of the Kiev Military District for some years and was very prominent in public affairs throughout his life. After Kovalevskaya was awarded the Prix Bordin of the Paris Academy of Sciences in December of 1888 and was already internationally recognized, Kosich used his influence with Grand Duke Konstantin Konstantinovich to facilitate her return to her homeland by having her elected to the Russian Academy of Sciences. She was denied full membership but was elected "Corresponding Member," an honor which carried with it no professorial position nor means of earning a livelihood in Russia.

8. There is an interesting section in a rough draft intended for the Swedish edition and later deleted, in which Kovalevskaya struggles to analyze her adolescent love for Dostoevsky. The passage begins and ends in mid-phrase:

> . . . in her imagination she elaborated and fleshed out many of those episodes in his life with other episodes which he had only mentioned in passing, and in her thoughts she relived them together with him. But she never thought about the future. The present was so beautiful, so rich and full. There is no doubt that if Dostoevsky could have looked into her soul and read her thoughts there, could have guessed, even by half, the depth of her feeling for him, he would have been touched by her boundless adoration. But it was not easy to see—that was the problem. To all appearances, Tanja was still a child.
>
> If Dostoevsky could have looked into Tanja's soul, he would undoubtedly have been deeply moved by what he saw there. But that is precisely the problem of the transitional phase of development that Tanja was going through: one's feelings run deep, almost like an adult's, but the way in which the feelings are expressed is

comical, childish, and it is hard for an adult to guess what is taking place inside the psyche of a fourteen-year-old girl.

Tanja understood Dostoevsky. Intuitively she comprehended the marvelous transports of tenderness buried in him. She venerated not only his genius, but also the suffering he had endured. Her own lonely childhood, her ever-present awareness that she was loved less than the other members of her family, had developed her inner world with much greater force than is usual with young girls of her years. From a very early age, she had felt a craving for some powerful, exclusive, all-encompassing attachment, and now, with the intensity which was the essence of her nature, she focused all her thoughts, all the capacity of her soul for ecstatic worship, on the first great man who crossed her path.

She thought incessantly about Dostoevsky. When she was alone she would review in her mind everything he had said during their last [every word of his, every random idea casually presented by him acquired a special significance in her eyes, she tried to understand the secret . . .] she endowed every word of his with profound meaning, strove to grasp [to penetrate], to elaborate every idea he had thrown out at random. What captivated her above all was precisely the originality of these ideas, the richness and novelty of the scenes and conjectures they conjured up in her imagination. There were also times when she abandoned herself to the wildest fantasies in connection with Dostoevsky. But strange to relate, these fantasies always concerned the past and not the future. Thus, for example, she would sit for hours on end and imagine herself at hard labor together with Dostoevsky. She relived in her th . . .

9. A very different report of the story was given by Dostoevsky in Anna Dostoevskaya's *Reminiscences,* according to which Dostoevsky had actually been engaged to Anyuta.

> When I asked why their marriage plans had come to nothing he replied, "Anna Vasilievna was one of the finest women I ever met in my life. . . . But her convictions were diametrically opposed to mine and she wasn't capable of giving in—she was too inflexible. And under such circumstances our marriage could hardly have been happy. I released her from her promise . . ." (op. cit., p. 56).

All available evidence suggests that Kovalevskaya's and not Dostoevsky's version is the correct one.

10. Kovalevskaya is mistaken on the date of Dostoevsky's engagement to Anna Grigorievna Snitkina. The Korvin-Krukovskys left Petersburg for Palibino in the spring of 1865. Dostoevsky's first meeting with the young woman who was to become his second wife took place on October 4, 1866. He proposed to her on November 8, and they were married on February 15, 1867. Both Sofya and Anyuta maintained a

warm continuing relationship with Dostoevsky and his wife after all of them were married, as their letters testify; and this despite the fact that the sisters had married political radicals of whom the Dostoevskys could not but strongly disapprove. Anyuta and her husband, the French Communist Victor Jaclard, even spent the summer of 1878 in Staraya Russia, where the Dostoevskys had a summer home. Of that summer Anna Dostoevsky remarks, "Almost every day after his walk my husband would go to have a talk with this fine, intelligent woman who had been so important in his life." (op. cit., p. 305)

Long after Dostoevsky's death Anna Dostoevskaya performed a very important service for Anyuta, who was then seriously ill. In March of 1887 Jaclard, suspected by the Russian police of underground political activity in connection with the attempted assassination of Alexander III, was ordered by the Minister of Internal Affairs to leave Russia within two days. This news brought Anyuta to the verge of emotional as well as physical collapse. Thereupon, Anna Dostoevskaya used her influence with the all-powerful Konstantin Pobedonostsev, Procurator of the Holy Synod, to have the deportation postponed for ten days, until such time as the Jaclards could pack their possessions and leave for Paris together. (Anyuta died in Paris that September.)

An Autobiographical Sketch[1]

As far as I can recall it now, my love of mathematics first showed itself in the following manner. I had an uncle, Pyotr Vasilievich Korvin-Krukovsky, my father's brother, who lived in his own village of Ryzhakovo, about twenty versts away from our estate. A man already in middle age, he had given over the complete charge of his estate to his only son,[2] and since he had a good deal of free time, he often came to visit us and lived in our house for months on end.

Uncle was an idealist in the full sense of the word, the type of person who, in many ways, is often termed "not of this world." He had been educated at home, and yet he possessed a vast and diversified store of knowledge, although (as is the case with most self-taught people), an inadequately grounded one. He had acquired it entirely through his own curiosity, with no outside help of any kind, and with only the most superficial basic preparation.

His favorite occupation, and the only real enjoyment left in his life, was reading. That was why our rural home library attracted him. He read without discrimination, and with equal pleasure, everything that came to his hands: novels, history, articles on popular science, scholarly treatises.

Endowed by nature with an uncommonly kind and gentle disposition, he loved children to distraction. Indeed, he himself had the soul of a child, with all that he was sixty years of age by then. And so, despite the difference in our years, Uncle and I developed a very close, almost chummy friendship.

I was fascinated by his stories while he, ever roving in a world of fantasies, often lost sight of the fact that it was a child he was with and, feeling the need to share his thoughts with someone, he would pour out his whole heart to me. I vividly remember the many long hours we spent together in the corner room of our big country house, the so-called "tower," which was also the library.

Uncle used to tell me fairytales and teach me how to play chess. Then, unexpectedly carried away by his own thoughts, he would initiate me into the secrets of the various economic and social projects through which he dreamed of benefiting humanity. But more than anything else, he loved to communicate the things he had succeeded in reading and learning in the course of his long life.

It was during such conversations that I first had occasion to hear about certain mathematical concepts which made a very powerful impression upon me. Uncle spoke about "squaring the circle," about the asymptote—that straight line which the curve constantly approaches without ever reaching it—and about many other things which were quite unintelligible to me and yet seemed mysterious and at the same time deeply attractive. And to all this, reinforcing even more strongly the impact of these mathematical terms, fate added another and quite accidental event.

Before our move to the country from Kaluga, the whole house was repainted and papered. The wallpaper had been ordered from Petersburg, but the quantity needed was not estimated quite accurately, so that paper was lacking for one room.

At first it was intended to order extra wallpaper from

Petersburg, but through rustic laxity and characteristic Russian inertia the whole matter was postponed indefinitely, as often happens in such situations. Meanwhile, time passed, and while everyone was intending and deciding and disposing, the redecoration of the rest of the house was finished.

Finally it was decided that it simply wasn't worth going to all the trouble of sending a special messenger to the capital, five hundred versts away, for a single roll of wallpaper. Considering that all the other rooms were in order, the nursery might very well manage without special paper. One could simply paste ordinary paper on the walls, particularly since our Palibino attic was filled with stacks of old newspapers accumulated over many years, lying there in total disuse.

By happy accident, however, it turned out that there in the attic, in the same pile with the old newspapers and other useless rubbish, were stored the lithographed lecture notes of Academician Ostrogradsky's[3] course on differential and integral calculus, which my father had attended as a very young Army officer. And it was these sheets which were utilized to paper the walls of my nursery.

I was then about eleven years old. As I looked at the nursery walls one day, I noticed that certain things were shown on them which I had already heard mentioned by Uncle. Since I was in any case quite electrified by the things he told me, I began scrutinizing the walls very attentively. It amused me to examine these sheets, yellowed by time, all speckled over with some kind of hieroglyphics whose meaning escaped me completely but which, I felt, must signify something very wise and interesting. And I would stand by the wall for hours on end, reading and rereading what was written there.

I have to admit that I could make no sense of any of it at all then, and yet something seemed to lure me on toward this occupation. As a result of my sustained scrutiny I learned many of the writings by heart, and some of the formulas (in their purely external form) stayed in my memory and left a deep trace there. I remember particularly that on the sheet of paper which happened to be on the most prominent place on the wall, there was an explanation of the concepts of infinitely small quantities and of limit. The depth of that impression was evidenced several years later, when I was taking lessons from Professor A. N. Strannolyubsky[4] in Petersburg. In explaining those very concepts he was astounded at the speed with which I assimilated them, and he said, "You have understood them as though you knew them in advance." And, in fact, much of the material had long been familiar to me from a formal standpoint.

It is to Yosif Ignatievich Malevich that I am indebted for my first systematic study of mathematics. It happened so long ago that I no longer remember his lessons at all: they remain a kind of dim recollection for me. But there is no question that they influenced me very much and were important in my later development.[5]

It was arithmetic which Malevich taught best and most innovatively. I have to confess, however, that when I first began to study with him, arithmetic held little interest for me. Probably because of Uncle Pyotr's influence, I was much more taken with various abstract considerations—infinity, for example. And, as a matter of fact, it is the philosophical aspect of mathematics which has attracted me all through my life. Mathematics has always seemed to me a science which opens up completely new horizons.

Besides arithmetic, Malevich taught me elementary geometry and algebra. Not until I grew somewhat more familiar with this latter field did I begin to feel an attraction to mathematics so intense that I started to neglect my other studies. Observing the direction I was taking, my father—who in any case harbored a strong prejudice against learned women—decided that it was high time to put a stop to my mathematics lessons with Malevich. But somehow I managed to wheedle out of my teacher a copy of Bourdon's *Algebra Course* and began studying it with diligence.

Since I was under my governess's strict surveillance all day long, I was forced to practice some cunning in this matter. At bedtime I used to put the book under my pillow and then, when everyone was asleep, I would read the night through under the dim light of the icon-lamp or the night lamp.

Under such circumstances, of course, I did not dare dream of continuing the systematic study of my favorite subject. My mathematical knowledge would likely have remained confined for a long time, to the contents of Bourdon's *Algebra* if I had not been aided by the following incident, which motivated my father to reassess his views on my education to some degree.

One of our neighboring landowners, Professor Tyrtov,[6] brought us the textbook of elementary physics he had written. I made an attempt to read it, but in the section on optics, to my chagrin, I encountered trigonometric formulas, sines, cosines and tangents.

What was a sine? I was nonplussed by this question and turned to Malevich for help in solving it. But since the matter was not a part of his curriculum, he replied that he did not know what a sine was.

Then, trying to cope with the formulas contained in the book, I tried to explain it for myself. By strange coincidence I took the same road that had been taken historically: that is, instead of a sine I used a chord. In the case of small angles these quantities almost coincided with one another. And since the formulas in Tyrtov's book dealt with infinitely small angles exclusively, they tallied excellently with the basic definition I had adopted. And with this I contented myself.

Some time later I was having a conversation with Professor Tyrtov about his book, and he expressed doubt at first that I could have understood it. To my declaration that I had read it with great interest he said, "Come, now—aren't you bragging?" But when I told him the means I had used to explain the trigonometric formulas he completely changed his tone. He went straight to my father, heatedly arguing the necessity of providing me with the most serious kind of instruction, and even comparing me to Pascal.

After some hesitation my father agreed to have me taught by Professor Strannolyubsky, with whom I then settled down to work successfully. In the course of that winter we went through analytic geometry, differential and integral calculus.

The next year I married V. O. Kovalevsky, and soon afterward my husband and I left for Europe, where we then went our different ways once more. I went to Heidelberg to continue with my mathematics studies, while my husband went on to another university to work in his own specialty of geology.

From Heidelberg I went on to Berlin, but there at the beginning I was faced with disillusionment. The capital of Prussia proved to be backward. Despite all my pleadings and efforts, I had no success in obtaining permission to attend the University of Berlin.

At that point Professor Karl Weierstrass took me under his wing. Thanks to my Heidelberg professors' reports on my work, and also because he saw that I was well prepared and motivated by a serious desire to work and not simply by the fashion then current, he suggested that we work together privately. These studies had the deepest possible influence on my entire career in mathematics. They determined finally and irrevocably the direction I was to follow in my later scientific work: all my work has been done precisely in the spirit of Weierstrass.

I consider Weierstrass himself one of the greatest mathematicians of all time, and indisputably the most outstanding of those alive today. He gave a completely new direction to the entire discipline of mathematics and created—not in Germany alone but in other countries as well—a whole school of young scholars who follow the road he marked out and develop his ideas.

At the same time that I was studying with Weierstrass I began to think of preparing for my doctorate. But since the doors of the University of Berlin were closed to me as a woman, I determined to try for Göttingen.

By German university regulations, doctoral requirements included, in addition to an examination, the presentation of a scholarly work, the so-called "inaugural dissertation." Weierstrass proposed several topics to me for further development. During the two years of my residence in Berlin, instead of the one dissertation required by regulations, I completed three: two in pure mathematics, "Zur Theorie der partiellen Differentialgleichungen" and "Über die Reduction einer bestimmten Klasse Abel'scher Integrale 3-en Ranges auf elliptische Integrale," and the third on a topic in astronomy, "Zusätze und Bermerkungen zu Laplace's Untersuchungen über die Gestalt der Saturnsringe."

All three works were presented to the University of Göttingen. They were adjudged sufficiently satisfactory for the university, contrary to its established procedure, to exempt me from the requirements of an examination and public defense of my dissertation (which is essentially no more than a formality) and to award me directly the degree of Doctor of Philoosphy, *summa cum laude.*

At the same time the first of the works mentioned above was published in *Crelle's Journal für die reine und angewandte Mathematik.* This honor, given to very few mathematicians, is particularly great for a novice in the field, inasmuch as *Crelle's Journal* was then regarded as the most serious mathematics publication in Germany. The best scientific minds of the day contributed to it, and such scholars as Abel and Jacobi had published their work in it in former times. My paper on astronomy, on the form of Saturn's rings, was not published until many years later, in 1885, in the journal *Astronomische Nachrichten.*

I returned to Russia in 1874. There I worked far less zealously than I had done in Germany and, indeed, the situation was far less propitious for scholarly work. My work was punctuated by long and frequent interruptions, so that I barely managed even to keep abreast of current developments in the field. In general, I did not complete a single independent study during this period of residence in Russia. The only thing that still gave me some feeling of scholarly support was the exchange of letters and ideas with my beloved teacher Weierstrass.

Various circumstances existed in Russia which distracted me from serious scientific work: society itself, and those conditions under which one had to live. At that time all of Russian society was pervaded by the spirit of profit-making and by the emergence of various commercial enterprises. This tendency took hold of my husband and, in part (I must confess my sins) of myself as well. We entered into the large-scale construction of stone houses with attached commercial bathhouses. But all this ended in failure and brought us to utter ruin.

Soon after my return to Russia, the newspaper *Novoye vremya* [New Times] began publication. My husband knew its publisher well, and in this way the two of us became a part of the *Novoye vremya* circle. It was in this newspaper that I first tested my literary ability in the capacity of theater critic.

In 1882 I went abroad again, and since that time I have lived there almost continuously, returning to Russia only rarely and for short periods on business. I have lived in many cities and in many countries during my life, so that I can say that I know Europe well, with the exception of Italy and Spain. Aside from Sweden, I know Paris best of all. I have been there many times, and even now I spend the greater part of my vacation holidays in France.

When I returned to Europe in 1882 I once again settled down seriously to the scientific work from which I had had so many years' respite in Russia. I went first to Paris, where I met the outstanding mathematicians of that time, including the renowned Hermite and also (of the younger mathematicians) Poincaré and Picard. These latter two, in my opinion, are the most gifted of the new generation of mathematicians in all of Europe.

It was at that time that I embarked on a major new work on the refraction of light in crystals. In the field of mathematics in general, it is mostly by reading the works of other scholars that one comes upon ideas for one's independent research. Thus I, too, was led to this topic by studying the work of the French physicist Lamé.

My work was completed in 1883 and had something of an impact in the mathematical world, for the problem of light refraction had not yet been satisfactorily clarified, and I had viewed it from a different, entirely new standpoint.

This paper of mine was published in 1884 in the new journal *Acta Mathematica,* which had come into being only two years earlier, in 1882. Although *Acta* is published in Sweden it is a truly international publication, inasmuch as it receives subsidies not only from the Swedish crown but also from foreign states including the government of France, as well as from Germany, Denmark and Finland. At this writing *Acta Mathematica* is regarded as one of the foremost mathematics journals in scholarly importance. Its contributors include the most distinguished scholars of all countries and deal with the most "burning" questions—those which above all others attract the attention of contemporary mathematicians.

It often happens that several persons will work on the same problem simultaneously. In general, publication conditions for a serious mathematics journal are quite different than for other periodicals. Therefore *Acta Mathematica* comes out not at prearranged intervals but in accordance with the accumulation of material, the development of new problems and the appearance of solutions to them. Normally, two volumes are published in the course of a year.

In addition to the paper on the refraction of light in crystals, several other articles of mine have appeared in *Acta Mathematica,* including (in 1883) the second of my doctoral dissertations, originally presented in 1874 at the University of Göttingen, (on the reduction of a certain class of Abelian functions to elliptic functions.)

All of my scholarly work is written in German or in French.[7] I am as much at home in them as I am in my native Russian. In mathematics, however, language plays a relatively non-essential role. Here the main thing is content, ideas, concepts, and to express these, mathematicians possess their own language: formulas.

Stockholm University in Sweden began to expand at the beginning of the 1880s, soon after its founding. At that time my name was fairly well known in the mathematics world, through my work and also through my acquaintance with almost all the eminent mathematicians of Europe. In Berlin and Paris I had especially frequent occasion to meet with the leading mathematician of Stockholm University, now its rector, Professor Gösta Mittag-Leffler, one of the most gifted students of our mutual teacher, Weierstrass.

And so in 1883 I was invited to Stockholm to lecture in mathematics. In this connection I should like to say a few words about the way this young university came into being.

Up to that time there had been a very old Swedish university, in existence for several hundred years, in the town of Uppsala. It suffered from the same flaws which can be found in almost all old universities located in small towns. In them the life seems, as it were, to have congealed: everything stays in the same form it had taken centuries before. The professors live their enclosed, almost medieval lives, a situation which contributes little to the development of new and fruitful ideas. Moreover one inevitably finds, just as in provincial Russian universities, a certain partiality, and professors engage in reciprocal favors.

In avoidance of this situation, the need began to be felt to summon forth fresh forces, and public opinion, which carries great weight in Sweden, began to demand the founding of a university to be located in the capital. Although Swedish life as a whole is characterized by great simplicity, still there are many wealthy people who willingly contribute substantial sums of money to social causes. Once society evinces sympathy toward a cause, it becomes easy to find the means to implement it. This fact cannot but strike every foreigner who comes to Sweden. One discovers that almost all of its institutions were built on private contributions.

So it was with the University of Stockholm as well. There was also another substantial reason which contributed to the desire to establish a new university: the inconvenience for many families residing in Stockholm of sending their young people to Uppsala, relatively far away.

The matter had a purely individual character at first. A few people came together and began of their common initiative to collect the necessary funds. But later, as public opinion began to express itself more forcefully, the government also—and mainly the municipality—decided to participate in the general cause and agreed to pay half of all the required costs. However, the government did not assume the role of decision-making for the future university; the question of the form in which it ought to develop was left to society itself.

From the very beginning, the basic principle laid down was that the university must be free. It was modeled on German universities in which teaching is conducted in complete freedom. [At Stockholm], for example, no credentials are required for attendance at lectures. Later, in order to obtain one's doctoral degree, one must present various kinds of documents, but one may attend lectures without obstacles of any kind, merely by paying an established fee. Moreover, women are permitted to attend lectures on a completely equal footing with men.

Examinations, everywhere invested with such importance, are not compulsory here. The university itself, up to the present writing, has yet to reach its full complement of departments (only two divisions are open) and until that time it does not possess the right to award scholarly degrees.

Teaching is not divided (as it is in Russia) into courses with predetermined curricula. Rather, the program is adapted to the quota of students. People often come to us to attend lectures and consult with professors (and such auditors are even more highly prized by us)—persons who have already studied in other institutions of higher learning, and even some who have already received scholarly degrees, from the University of Uppsala, for instance, or from the University of Lund. Many young people also come from Finland. Some of these are doctoral candidates at the University of Helsingfors, and some of our finest students have been Finnish.

The school year is divided into two semesters separated by holiday periods. Shortly before the beginning of each semester we professors meet to discuss the courses to be organized for the forthcoming semester. At the present time Stockholm University numbers more than two hundred students, both men and women. Last year about one-third of the student body belonged to the best families in Sweden. In general, our university youth are splendid young people, and the students' relationships, both with one another and with the faculty, are of the warmest.

Many of our former students have already embarked upon independent careers. Two, for example, have become docents at the University of Helsingfors, and one is a lecturer at the Higher Technical Institute. One of our women students is a member of the faculty for the senior classes of a boys' *gymnasium,* and she now teaches fifteen- and sixteen-year-old boys.

When I first came to Sweden I offered the alternatives of lecturing either in German or in French. The majority of auditors preferred that I lecture in German. But after the first year I was already able to lecture in Swedish. This presented no particular difficulties for me because I came into social contact with Swedes immediately upon my arrival, and began to take lessons in the Swedish language.

The position offered me at the beginning was that of docent.[8] But in less than a year I was appointed staff professor, a position I have held since 1884. In addition to lecturing, my duties also include participation at university faculty meetings, and I have the right to vote on an equal basis with the other professors. The salary of our staff professors comes to about 6000 kronor a year (the Swedish krona is slightly higher than the German mark, and 700 kronor are equal to 1000 French francs). I give four lectures a week, that is, a two-hour session for two days in succession. Since the content of my course deals with extremely specialized problems, my auditors are not very many: seventeen or eighteen persons.

During my first year of residence in Sweden I worked very hard and very seriously. Among other things, I wrote there the most important of my mathematical works, for which I was awarded a prize by the Paris Academy of Sciences. This work examined the problem of the rotation of a solid body around a fixed point under the influence of gravitational force. The problem, which encompasses the theory of the pendulum, is a highly significant one. It is also one of the classical problems of mathematics; some of the greatest minds, including Euler, Lagrange and Poisson, have applied their efforts to its solution.

And yet the problem is still far from being completely

solved. We know of only a few cases for which a complete and rigorous mathematical solution has been found. In the whole history of mathematics there are not many problems like this one—in such compelling need of solution, to which so much intellectual power and stubborn labor have been applied without leading to substantive results in the majority of cases. Not for nothing have German mathematicians dubbed the problem *die mathematische Nixe.*

The problem had always held a strong interest for me. I began work on it long ago, almost from my student days. But my efforts remained fruitless for a long time; not until 1888 were they crowned with success. Therefore, my happiness can be imagined when I succeeded at long last in achieving a really substantial result and in taking an important stride forward toward the solution of so difficult a problem.

In that same year of 1888, the Paris Academy of Sciences announced a prize competition for the best essay "Sur le problème de la rotation d'un corps solide autour d'un point fixe," with the proviso that the essay must substantially refine or supplement findings previously attained in this area of mechanics.

At that time I had already achieved the main results of my work. But they still existed only inside my own head. And since the problem I had solved was entirely appropriate to the topic assigned by the Paris Academy, I set to work with even greater zeal to put all my material in order, to work out the details and write the paper by the designated deadline.

When all of this had been satisfactorily completed I sent my manuscript to Paris. By the rules of the competition, it was to be submitted anonymously. Therefore, I noted down a maxim[9] on the manuscript and attached to it a sealed envelope containing my name, with the same maxim inscribed above it. This was the procedure by which authors retained their anonymity while their work was being evaluated.

The result was beyond what I had hoped. Some fifteen papers were present, but it was mine which was found deserving of the prize. And that was not all: in view of the fact that the same topic had been assigned three times running and had remained unsolved each time, and also in view of the significance of the results achieved, the Academy voted to increase the previously announced award of 3000 francs to 5000.

The envelope was then unsealed, and it was learned that I was the author of the work. I was informed immediately and left for Paris to be present at the session of the Academy of Sciences set for the occasion. I was given a highly ceremonial reception and seated next to the president, who made a flattering speech; all in all, honors were lavished upon me.

As I have already mentioned, I have been living in Sweden since 1883 and have become so assimilated to its style of life that I feel completely at home there. Stockholm is a very beautiful city and its climate is rather good, except for the spring weather, which is unpleasant. I have a large circle of friends and lead an active social life. I am even received at Court.

With respect to Swedish society I have to say that its educated stratum differs very little from that of Petersburg. But if one considers the social body as a whole, then the level of Swedish society, of course, is immeasurably higher than in Russia.

A distinguishing trait of the Swedish people is their uncommon good nature and mildness, characteristics which developed in them, I believe, because their history has never been a history of oppression. Political parties do exist, of course, but their struggle takes more moderate forms—one is not so aware, somehow, of the desire to spite each other. The clearest manifestation of the national character is shown in the relations between Sweden and Norway. In most cases the former makes concessions to the latter, so that Norway, which was united with Sweden at one time, now enjoys equal rights in every respect.

King Oscar is a pleasant and cultivated person. As a young man he attended lectures at the university, and still today shows an interest in science, although I cannot vouch for the profundity of his erudition. He has no official contact with the university but is extremely sympathetic to it and very amicably disposed toward its professors in general and to myself in particular. In the political sphere he manifests those qualities common to all the Swedes: moderation and willingness to compromise. Thus, for example, where there was some dissension between the King and his ministers in certain areas during the last so-called "Ministerial Affair," it was the King who ended by making the concessions.

S. V. KOVALEVSKAYA,
née Korvin-Krukovskaya
St. Petersburg
May 29, 1890

Notes to an Autobiographical Sketch

1. First published in *Russkaya starina,* No. 1, 1891, 450–463.
2. See Chapter Six, note 1, p. 203.
3. Ibid., note 5.
4. Ibid., note 8.
5. This shift in Kovalevskaya's assessment of Malevich's influence on her intellectual development (see Chapter 3, 79: " . . . although his actual influence on my education was slight") was probably occasioned by the fact that the editor of *Russkaya starina,* M. I. Semevsky, had been a pupil of Malevich's in his own childhood. Malevich himself had been highly incensed by Kovalevskaya's slighting remark, and in response he wrote a lengthy reminiscence of his nine years as tutor in the Korvin-Krukovsky family, elaborating his entire course of study in detail. This reminiscence was published in *Russkaya starina* in 1890, except for a very bitter section (unpublished) in which Malevich depreciated the influence of Uncle Pyotr Vasilievich and Professor Strannolyubsky, and maintained that it was he who had started Kovalevskaya on her mathematical career.
6. N. N. Tyrtov, Professor of Physics at the Naval Academy.
7. Not until 1948 were Kovalevskaya's complete scientific works translated into Russian and collected for publication (S. V. Kovalevskaya, *Nauchnye raboty,* edited by P. Ya. Kochina, trans. by P. Ya. Kochina, L. A. Tesheva, T. O. Levina, Moscow, 1948).
8. Not quite accurate. The position offered Kovalevskaya was privat-docent, roughly equivalent to assistant professor, but without official staff affiliation, and with salary to be paid by the students directly to the teacher. This was a normal starting point at European universities for young beginning scholars, but was far below what might have been expected in Kovalevskaya's case.
9. *Die ce que tu sais, fais ce que tu dois, adviendra que pourra!* (Say what you know, do what you must, and whatever will be, will be.)

On the Scientific Work of Sofya Kovalevskaya

On the Scientific Work of Sofya Kovalevskaya

by

P. Y. Polubarinova-Kochina

During her lifetime, Sofya Kovalevskaya published ten articles in mathematics and physics, [1–10] two of which [5, 6] are identical except for the languages in which they were published. Not until the Soviet period were all her papers translated into Russian and collected in book form [11]. We shall dwell here in detail on the two basic themes with which Kovalevskaya was concerned.

The first of these is the Cauchy-Kovalevsky Theorem. In 1874, Weierstrass presented to the University of Göttingen three doctoral dissertations written by Kovalevskaya, of which the first, "Zur Theorie der partiellen Differentialgleichungen," was the most important. This paper is now regarded as the first significant result in the general theory of partial differential equations. The theorem contained in it is known as the Cauchy-Kovalevsky Theorem.

In 1842, the French mathematician Augustin Cauchy had begun a series of four publications on integrating differential equations with initial conditions. He proved the existence of analytic solutions of a problem which acquired the name of "Cauchy's problem" for ordinary differential equations and for a certain class of linear partial differential equations. He also showed how a nonlinear system can be reduced to this case.

For the ordinary differential equation

$$\frac{dy}{dx} = f(x,y) \tag{1}$$

with the initial condition

$$y = y_0 \quad \text{and} \quad x = x_0. \tag{2}$$

Cauchy's problem is formulated as follows:

If $f(x, y)$ is an analytic function in a neighborhood of the point (x_0, y_0), i.e., if it can be expanded in a series of non-negative integer powers of $x - x_0$, $y - y_0$, there exists a unique solution $y(x)$ of equation (1) with condition (2).

The proof of this and other analogous theorems of Cauchy was carried out by the method of majorants. For problem (1), (2) this method consists of replacing $f(x, y)$ by the majorant $F(x, y)$, i.e., by a simple analytic function whose power series expansion coefficients are non-negative and greater than or equal to the absolute value of the corresponding series coefficients for $f(x, y)$. Then the equation

$$\frac{dy}{dx} = F(x,y), \qquad y(0) = a \qquad (a = y_0)$$

is integrated explicitly and the corresponding solution is a majorant for the solution to problem (1), (2) (with $x_0 = 0$). The majorant constructed by Cauchy had the form

$$\frac{M}{\left(1 - \frac{x}{a}\right)\left(1 - \frac{y}{b}\right)}, \tag{3}$$

where M, a,b are constants.

Like Weierstrass, Kovalevskaya set herself the problem

without knowing Cauchy's work. In Weierstrass's lectures on the theory of ordinary differential equations, majorants of the form

$$\frac{M}{1 - \dfrac{x + y}{a}}, \tag{4}$$

where M, a are constants, were used, which are simpler for computations than expression (3).

Kovalevskaya proved an existence theorem for an analytic solution satisfying given initial conditions, at first for a quasi-linear system of partial differential equations, and later for a general non-linear system of arbitrary order in *normal form,* by reducing it to a quasi-linear system. In doing this Kovalevskaya used majorants of the form (4).

In order to formulate Kovalevskaya's theorem, we shall use Petrovsky's terminology [12]. Consider the system of equations

$$\frac{\partial^{n_i} u_i}{\partial t^{n_i}} = F_i\left(t, x_1, \ldots, x_n, u_1, \ldots, u_N, \ldots, \frac{\partial^k u_j}{\partial t^k \partial x_1^{k_1} \ldots \partial x_n^{k_n}}, \ldots\right) \tag{5}$$

$$(i, j = 1, 2, \ldots, N;\ k_0 + k_1 + \ldots + k_n = k \leqslant n_j,\ k_0 < n_j).$$

We see that for each unknown function u_i there exists a highest order n_i of the derivatives of this function which appear in the system. The derivative $\partial^{n_i} u_i/\partial t^{n_i}$ must be among the derivatives with respect to t of the highest order n_i of each function u_i which appears in the system; the system is then solved for these derivatives *(normal form).*

At some point $t = t_0$ we are given initial values of the unknown functions u_i and their first $n_i - 1$ derivatives with respect to t:

$$(6) \qquad \left(\frac{\partial^k u_i}{\partial t^k}\right)_{t=t_0} = \varphi_i^{(k)}(x_1, x_2, \ldots, x_n),$$

$$(k = 0, 1, \ldots, n-1)$$

(when $k = 0$, we have the function u_i itself).

All the functions $\varphi_i^{(k)}$ are defined in the same region $G(x_1, \ldots, x_n)$.

Cauchy's problem consists of finding a solution of system (5) with initial conditions (6).

Now, Kovalevskaya's theorem is formulated as follows:

If all the functions F_i are analytic in a certain neighborhood of the point $(t^0, x_1^0, \ldots, x_n^0, \ldots, \varphi^0_{j,k_0,k_1,\ldots,k_n}, \ldots)$ and all the functions $\varphi_j^{(k)}$ are analytic in a neighborhood of the point $(x_1^0, \ldots, x_n^0)$, then Cauchy's problem has an analytic solution in a certain neighborhood of the point $(t^0, x_1^0, \ldots, x_n^0)$, and, moreover, it is the unique solution in the class of analytic functions. In this connection, $\varphi^0_{j,k_0,k_1,\ldots,k_n} = (\partial^{k-k_0} \varphi_i{}^{(k_0)}/\partial x_i^{k_1} \ldots \partial x_n^{k_n}\, x_i = x_i^0)$.

Note: A function $F(z_1, \ldots, z_m)$ is called analytic in a neighborhood of the point $(z_1^0, \ldots, z_m^0)$ if it can be expanded in the power series

$$F(z_1, \ldots, z_m) = \sum_{k_1 \ldots k_m} A_{k_1 \ldots k_m}(z_1 - z_1^0)^{k_1} \ldots (z_m - z_m^0)^{k_m}$$

When the Cauchy-Kovalevsky Theorem is discussed, usually Poincaré's statement is cited: "Kovalevskaya significantly simplified the proof and gave the theorem its definitive form."

While working on the theorem, Kovalevskaya first examined the simplest equation of heat conduction

$$\frac{\partial \varphi}{\partial t} = \frac{\partial^2 \varphi}{\partial x^2}. \tag{7}$$

She sought a solution satisfying the condition $\varphi(x, t) = \varphi_0(x)$ when $t = 0$. In the special case $\varphi_0(x) = x/(1 - x)$ (for $|x| < 1$) it turned out that a solution, if it existed, would give the power series in t:

$$\sum_{n=0}^{\infty} \frac{(2n)!}{n!} \frac{t^n}{(1 - x)^{2n+1}} \tag{8}$$

which, however, diverges at any point if $t \neq 0$, i.e., the problem has no analytic solution—equation (7) is not written in normal form.

Weierstrass (like Poincaré at a later date) greatly valued this result. In a letter to Kovalevskaya of May 6, 1874, he discusses the problem of representing the solution of equation (7) with $\varphi(x, 0) = \varphi_0(x)$ as an integral, and adds:

> "So you see, dear Sonya, that your observation (which seemed so simple to you) on the distinctive property of partial differential equations, namely, that an infinite series may *formally* satisfy such a differential equation and nonetheless not converge for any system of values of the variables, was for me the starting point for interesting and very elucidating researches. I would like my student to go on expressing her gratitude to her teacher and friend in the same way.
>
> With warmest regards,
> Yours, K. WEIERSTRASS."

In recent times, other methods besides the method of majorants have been used to prove the Cauchy-Kovalevsky Theorem and its generalizations, such as the method of successive approximations, and the method of reduction to a symmetrical system (cf. Courant Hilbeit *Methods of Mathematical Physics*).

As an example of the scope of interest in Kovalevskaya's theorem, one might cite Mizohata's article in *Uspekhi matematicheskikh nauk,* "On Kovalevskian Systems," in which the author uses concepts associated with the name of Kovalevskaya, such as Kovalevskian systems, Kovalevskian polynomials. Leray and others have made interesting generalizations.

At the present time, the theory of partial differential equations has attained a high level of development. Petrovsky identified the classes of hyperbolic, parabolic and elliptic systems which were subjected to comprehensive study by him and, in later works, by a number of other mathematicians as well. The qualitative properties of solutions have been studied; the conditions necessary for the correct formulation of problems have been clarified. During the past twenty years the following problem, posed by Petrovsky, has been studied: does a non-analytic equation have even a single solution? In this connection, Kovalevskaya's theorem and its place in the general theory of partial differential equations have become very apparent [13]. It has been shown that the slightest deviation from the conditions of the theorem lead to unsolvable equations. The first example of this type was given by Lewy, who showed that the equation

$$\frac{1}{2}(u_x + iu_y) - i(x + iy)u_t = f(x, y, t)$$

in general, has no solutions for "most" infinitely differentiable functions $f(x, y, t)$.

In her talk, "Kovalevskaya's Theorem and its rôle in the modern theory of partial differential equations," presented at the Institute for Problems in Mechanics of the Academy of Sciences on the occasion of the 125th. anniversary of Kovalevskaya's birth, Oleinik said, "Kovalevskaya's Theorem has found important and substantial applications in research on the theory of partial differential equations carried out up to the present day, and much of the most sophisticated contemporary research bears witness to its profound and definitive character."

The well-known problem which made Kovalevskaya famous was the problem of the motion of a heavy rigid body near a fixed point. Examples of this phenomenon are the movement of the pendulum and, under certain assumptions, the motion of a top and of a gyroscope.

One considers a heavy rigid body rotating around a fixed point which is taken as the origin of two coordinate systems: the fixed system (X, Y, Z) with the Z-axis vertical, and the moving system (x, y, z) which is fixed relative to the body (see fig. 1). The axes of the latter system are directed along the principal axes of the ellipsoid of inertia of the body. The principal moments of inertia of the body are

$$A = \int_\tau (y^2 + z^2)\, \rho\, d\tau,$$

$$B = \int_\tau (x^2 + z^2)\, \rho\, d\tau, \qquad C = \int_\tau (x^2 + y^2) \rho\, d\tau,$$

where τ is the volume of the body, $d\tau = dx\, dy\, dz$, and ρ its density.

Let $\gamma, \gamma', \gamma''$ be the directional cosines of the Z-axis relative to the moving axes x, y, z (fig. 1). Let M denote the mass of the body, (x_0, y_0, z_0) the coordinates of its center of gravity, and g the acceleration due to gravity.

At an arbitrary point (x, y, z) of the body, the velocity vector $\vec{v}$ is expressed in terms of the angular velocity vector $\vec{\Omega}$ and the radius vector $\vec{r}(x, y, z)$ of the point as

$$\vec{v} = \vec{\Omega} \times \vec{r},$$

with the components of the vector $\vec{\Omega}$ being denoted by p, q, r.

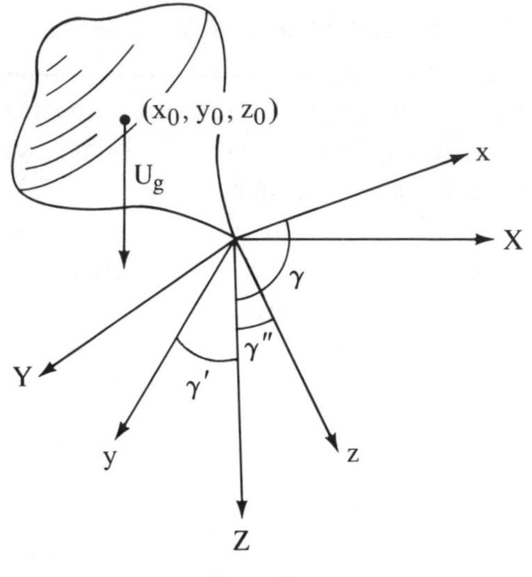

Figure 1

The first group of equations of the motion of a heavy rigid body having a fixed point, which were derived by Euler in 1750, are

$$(9)\qquad \begin{aligned} A\frac{dp}{dt} + (C - B)qr &= Mg(y_0\gamma'' - z_0\gamma'),\\ B\frac{dq}{dt} + (A - C)rp &= Mg(z_0\gamma - x_0\gamma''),\\ C\frac{dr}{dt} + (B - A)pq &= Mg(x_0\gamma' - y_0\gamma). \end{aligned}$$

The second group of equations of motion gives the connection between p, q, r and γ, γ', γ'':

$$(10)\qquad \begin{aligned} \frac{d\gamma}{dt} &= r\gamma' - q\gamma'',\\ \frac{d\gamma'}{dt} &= p\gamma'' - r\gamma,\\ \frac{d\gamma''}{dt} &= q\gamma - p\gamma'. \end{aligned}$$

The problem consists of finding $p(t)$, $q(t)$, $r(t)$, $\gamma(t)$, $\gamma'(t)$, $\gamma''(t)$ if one knows the values of the functions $p_0, q_0, r_0, \gamma_0, \gamma'_0, \gamma''_0$ at some initial time, i.e., in solving Cauchy's problem for the system (9), (10). For geometric reasons, the relation of $\gamma^2 + \gamma'^2 + \gamma'^2 = 1$ must be fulfilled.

The system of equations (9), (10), containing six unknown functions, has the three first integrals

$$
\begin{aligned}
&Ap^2 + Bq^2 + Cr^2 - 2Mg(x_0\gamma + y_0\gamma' + z_0\gamma'') = C_1 = h,\\
(11)\qquad &Ap\gamma + Bq\gamma' + Cr\gamma'' = C_2 = \mathrm{k},\\
&\gamma^2 + \gamma'^2 + \gamma''^2 = C_3 = 1.
\end{aligned}
$$

It might be thought that three additional integrals would be necessary for a complete solution of the problem. However, the fact that time enters the equations only in the form dt permits us, after dividing all the equations term-by-term by one of the equations, to eliminate time and obtain five equations. Further, the use of an integrating factor permits us to find one additional integral. Therefore, once we have (11), it is sufficient to find one more, fourth, integral to obtain a complete solution of the problem in quadratures. Special cases were known in which there was a fourth integral.

(1) Euler's case (1750–1785). Here we take $x_0 = y_0 = z_0 = 0$, i.e., the centre of gravity is at the fixed point. Then if we take equations (9), multiply them successively by Ap, Bq, Cr, add them, and integrate, we find a fourth integral:

$$A^2p^2 + B^2q^2 + C^2r^2 = C_4 = l^2.$$

After several algebraic transformations using equations (9) we obtain a solution, of which we shall give only the expression for q when $B > D$:

$$n(t - t_0) = -\int_0^u \frac{du_1}{(1 - u_1^2)^{1/2}(1 - x^2u_1^2)^{1/2}} = -F(u, x), \qquad (12)$$

where we take

$$A > B > C,$$
$$u = Nq,$$
$$N = (l/h)\{[B(B - C)]/[D(D - C)]\}^{1/2},$$
$$D = l^2/h$$
$$x^2 = [(A - B)(D - C)]/[(A - D)(B - C)],$$
$$n = [(A - B)(B - C)/AC]^{1/2}.$$

We find q from equation (12) by inverting the elliptic integral $F(u, x)$:

$$q = u/N = -(1/n)\mathrm{Sn}(n(t - t_0), \tag{13}$$

where Sn is the elliptic sine.

We similarly determine p and r, where we must consider different ranges of values of D.

The following relations are found for $\gamma, \gamma', \gamma''$:

$$\gamma = Ap/l, \qquad \gamma' = Bq/l, \qquad \gamma'' = Cr/l.$$

Poinsot (1851) investigated this case from a geometric point of view.

(2) Lagrange's case (1788). Here we take $A = B$, $x_0 = y_0 = 0$, i.e., we consider a body with a symmetric ellipsoid of inertia and with its center of gravity on the z-axis. The last equation in (9) takes the form $C dz/dt = 0$, i.e., $r = C_4 = R$ is a new fourth, integral. Here, solving the problem again reduces to inverting elliptic integrals. Lagrange's investigations were later developed by Poisson (1813). The expression for the general integrals of the system is due to Jacobi.

(3) $A = B = C$. This is the case of complete kinetic symmetry, when the ellipsoid of inertia is a sphere. Here, one has a fourth algebraic integral:

$$x_0 p + y_0 q + z_0 r = C_4.$$

For a long time after these investigations, no essential progress was made in the problem of a rotating solid.

Kovalevskaya become interested in this problem, which was called the "mathematical mermaid" because of its attractiveness and elusiveness. In a letter to Mittag-Leffler of November 21, 1881, Kovalevskaya writes that she was doing research in optics, but "I have a hard time tearing myself away from work on another question, which has been on my mind almost from the very beginning of my mathematical studies. . . . It concerns the solution of the general case of the problem of a heavy solid rotating around a fixed point using Abelian functions. Weierstrass once suggested that I take up this problem, but at that time all my attempts were fruitless, and even the work of Weierstrass himself showed that the differential equations of this problem cannot be satisfied by single-valued functions of time. His result then made me give up trying to solve this problem. But since then, the excellent, still unpublished research of our teacher on stability conditions and the analogy with other dynamical problems has rekindled my enthusiasm and inspired me with the hope of solving this problem using Abelian functions *whose arguments are not linear functions of time.* This research struck me as so interesting and engaging that for the time being I have forgotten everything else and given myself over to this work with all the fervor of which I am capable."

Kovalevskaya adopted a new approach to the rotation problem. She considered time as a complex variable and applied the theory of functions of a complex variable to the problem in mechanics. She asked the question: When can a solution be found in the form of series of integral powers of $t - t_0$ (t_0 is a complex number) including a certain number of negative powers, i.e., one supposes that the integrals of the equations of motion have poles on the complex plane? With $\tau = t - t_0$:

$$
(14) \qquad
\begin{aligned}
p &= \tau^{-m_1}(p_0 + p_1\tau + p_2\tau^2 + \ \ldots),\\
\gamma &= \tau^{-n_1}(f_0 + f_1\tau + \ldots),\\
q &= \tau^{-m_2}(q_0 + q_1\tau + q_2\tau^2 + \ldots),\\
\gamma' &= \tau^{-n_2}(g_0 + g_2\tau + \ldots),\\
r &= \tau^{-m_3}(r_0 + r_1\tau + r_2\tau^2 + \ldots),\\
\gamma'' &= \tau^{-n_3}(h_0 + h_1\tau + \ldots).
\end{aligned}
$$

Substituting these series in equations (9), (10) made it possible to determine the order of the poles ($m_1 = m_2 = m_3 = 1$, $n_1 = n_2 = n_3 = 2$) and obtain the required relations between the coefficients of the equations. It turned out that solutions of the form (14) are possible in only four cases:

(1) $x_0 = y_0 = z_0 = 0$ (the Euler-Poinsot case);
(2) $A = B$, $x_0 = y_0 = 0$ (the Lagrange-Poisson case);
(3) $A = B = C$ (the case of complete kinetic symmetry);
(4) $A = B = 2C$, $z_0 = 0$ (the Kovalevskaya case).

Figure 2 shows three tops, drawn by Zhukovsky to correspond to the cases of Euler, Lagrange and Kovalevskaya.

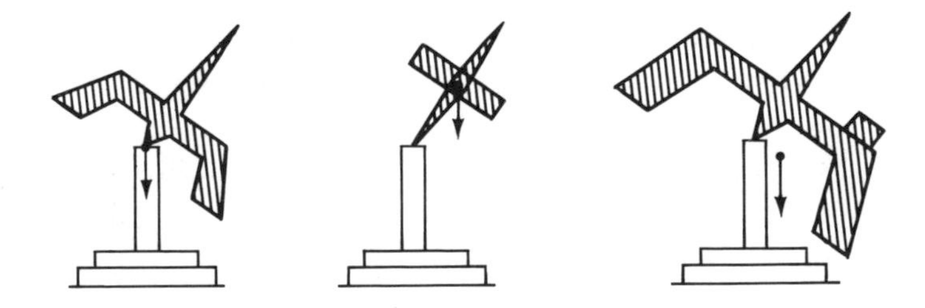

For her own case, Kovalevskaya found a fourth algebraic integral:

$$
(15) \quad (p^2 - q^2 - c\gamma)^2 + (2pq - c\gamma') = C_4 = k^2, \qquad \left(c = \frac{Mgx_0}{C}\right).
$$

Consequently, all the integrals of motion could be found after reducing to quadratures and using the last factor theorem.

However, in the case discovered by Kovalevskaya, the reduction to quadratures of the equations of motion turned out to be incomparably more difficult than in the three earlier cases. I shall give only several of the steps in her computations.

In the Kovalevskaya case, with $c = Mgx_0/C$, $y_0 = 0$, equations (9) are:

$$2\frac{dp}{dt} = qr, \qquad 2\frac{dq}{dt} = -pr - c\gamma'', \qquad \frac{dr}{dt} = c\gamma'. \tag{16}$$

We add equations (10) to these equations.

The four first integrals can be written in the form

$$\begin{aligned} &2(p^2 + q^2) + r^2 = 2c\gamma + 6l_1, \\ &2(p + q\gamma') + r\gamma'' = 2l, \\ &\gamma^2 + \gamma'^2 + \gamma''^2 = 1, \\ &[(p + qi)^2 + c(\gamma + i\gamma')][(p - qi)^2 \\ &+ c(\gamma - i\gamma')] = k^2. \end{aligned} \tag{17}$$

Kovalevskaya's first transformation is to introduce the new functions x_1, x_2, ζ_1, ζ_2 in place of p, q, γ, γ':

$$\begin{aligned} &x_1 = p + qi, \qquad x_2 = p - qi, \\ &\zeta_1 = (p + qi)^2 + c(\gamma + i\gamma') = x_1^2 + c(\gamma + i\gamma'), \\ &\zeta_2 = (p - qi)^2 + c(\gamma - i\gamma') = x_2^2 + c(\gamma - i\gamma'). \end{aligned} \tag{18}$$

In terms of these variables the four first integrals (17) take the form:

$$
(19)\qquad
\begin{aligned}
r^2 &= E + \zeta_1 + \zeta_2,\\
cr\gamma' &= F - x_2\zeta_1 - x_1\zeta_2,\\
c^2\gamma'' &= G + x_2^2\zeta_1 + x_1^2\zeta_2,\\
\zeta_1\zeta_2 &= k^2,
\end{aligned}
$$

where E, F, G are polynomials in x_1, x_2:

$$
(20)\qquad
\begin{aligned}
E &= 6l_1 - (x_1 + x_2)^2,\\
F &= 2cl + x_1x_2(x_1 + x_2),\\
G &= c^2 - k^2 - x_1^2x_2^2.
\end{aligned}
$$

Here, $l_1\, l,\quad ,k$ are arbitrary constants.

r and γ'' are eliminated from the first three equations in (19). Then ζ_1 and$l\gamma_2$ can be expressed in terms of x_1 and x_2, and we obtain equations containing x_1 and x_2.

We introduce the following polynomials in x_1, x_2:

$$
\begin{aligned}
R(x_s) &= Ex_s^2 + 2Fx_s + G \qquad (s = 1, 2),\\
R(x_1,x_2) &= Ex_1x_2 + F(x_1 + x_2) + G,
\end{aligned}
$$

and let s_1 and s_2 denote the expressions:

$$
(21)\qquad
\begin{aligned}
s_1 &= \frac{R(x_1, x_2) - [R(x_1)R(x_2)]^{1/2}}{(x_1 - x_2)^2} + 3l_1,\\
s_2 &= \frac{R(x_1, x_2) + [R(x_1)R(x_2)]^{1/2}}{(x_1 - x_2)^2} + 3l_1.
\end{aligned}
$$

Then, using the equations for dx_1/dt and dx_2/dt, one eventually obtains the following relations:

$$
(22)\qquad
\begin{aligned}
\frac{dx_1}{[R(x_1)]^{1/2}} + \frac{dx_2}{[R(x_2)]^{1/2}} &= -i\,\frac{[(s_2 - l_4)(s_2 - l_5)]^{1/2}}{s_1 - s_2},\\
\frac{dx_1}{[R(x_1)]^{1/2}} + \frac{dx_2}{[R(x_2)]^{1/2}} &= i\,\frac{[(s_1 - l_4)(s_1 - l_5)]^{1/2}}{s_1 - s_2},
\end{aligned}
$$

with $l_4 = 3l_1\text{-}k$ and $l_5 = 3l_1 + k$.

Next, using the addition theorem for elliptic functions, one proves that the left-hand sides in (22) can be expressed in terms of the variables s_1 and s_2. We finally obtain the following equations for s_1 and s_2:

$$\frac{ds_1}{[\varphi(s_1)]^{1/2}} + \frac{ds_2}{[\varphi(s_2)]^{1/2}} = 0,$$
$$\frac{s_1 ds_1}{[\varphi(s_1)]^{1/2}} + \frac{s_2 ds_2}{[\varphi(s_2)]^{1/2}} = \frac{1}{2}\, i\, dt.$$

Suppose that when $t = t_0$ (now t_0, the initial time, is real) we have $s_1 = s_{10}$, $s_2 = s_{20}$. We can then write

$$\begin{aligned} &\int_{s_{10}}^{s_1} \frac{ds_1}{[\varphi(s_1)]^{1/2}} + \int_{s_{20}}^{s_2} \frac{ds_2}{[\varphi(s_2)]^{1/2}} = 0, \\ &\int_{s_{10}}^{s_1} \frac{s_1 ds_1}{[\varphi(s_1)]^{1/2}} + \int_{s_{20}}^{s_2} \frac{s_2 ds_2}{[\varphi(s_2)]^{1/2}} = \frac{1}{2}\, i\,(t - t_0). \end{aligned} \tag{23}$$

Here $\varphi(s)$ is a fifth degree polynomial

$$\varphi(s) = (s - l_1)(s - l_2)(s - l_3)(s - l_4)(s - l_5)$$

having roots l_s $(s = 1, 2, 3, 4, 5)$, whose values we shall not write down, except for the values given above: $l_4 = 3l_1 - k$ and $l_5 = 3l_1 + k$. The integrals, which contain rational functions in x and $[\varphi(x)]^{1/2}$, where $\varphi(x)$ is a polynomial of degree greater than four, are called hyperelliptic integrals. Thus, it is now necessary to solve the problem of inverting the hyperelliptic integrals in equations (23) in order to find $s_1(t)$ and $s_2(t)$. It will then be possible to find x_1 and x_2 from equations (21) and, as the final result, use (19) to find $p, q, r, \gamma, \gamma', \gamma''$ as functions of time.

The very elegant system (23) is thereby obtained by what Golubev [14] has called the virtuoso transformations of Kovalevskaya. The rest of the problem—inverting the resulting integrals—is extremely complicated. Here Kovalevskaya, with great skill, applied the theta-functions of Jacobi, which are connected with elliptic functions. Successive authors have attempted to simplify the computations [14].

The special cases where $\phi(s)$ has a double or triple root were considered by Appelrot, Delone and Mlodzevsky.

In these cases, the integrals in (23) become elliptic integrals or elementary integrals.

It should also be noted that Zhukovsky gave a geometric interpretation of Kovalevskaya's solution, and Delone constructed a Kovalevskian gyroscope apparatus. Golubev [14] has given a deep analysis of Kovalevskaya's research from the point of view of the theory of functions of a complex variable.

Kovaleskaya continued to think about the general solution to the problem of a rotating heavy solid. In a letter to Mittag-Leffler, written in the autumn of 1888, Kovalevskaya writes: "I am describing in a letter to Hermite what seem to me to be surprising and interesting results which I have found relating to the general case." However, after Kovalevskaya's death, no additional material on the rotation problem was found among her papers.

Poincaré proved that, under certain general conditions, the equations of dynamics do not have any algebraic first integrals except for the classical ones. Research by Liouville, Husson and Burgatti showed that the rotation problem has an algebraic first integral only in the cases noted by Kovalevskaya.

Lyapunov (1894) used the method of variations to show that one can obtain the cases distinguished by Kovalevskaya without limiting the problem to finding solutions with poles; it is sufficient to require single-valued integrals—for example, there can be essential singularities, or there may be no singularities except at infinity.

The value of Kovalevskaya's work is not only in the results themselves nor in the originality of her method, but also in the increased interest she aroused in the problem of a rotating heavy solid on the part of researchers in many countries, in particular in Russia. The results of this research for the most part consist of finding new cases of *special* solutions of the problem, when not all of the arbitrary constants C_1, C_2 . . . remain arbitrary. Examples are the work of Hessa, Chaplygin, Bobylev, Steklov, Goryachev, N. Kowalevsky, and Zhukovsky. In recent years, many cases have been found and studied by P. V. Kharlamov and E. I. Kharlamova and their students. Kharlamov gives geometrical interpretations of the motions, including Kovalevskaya's case [16].

Kovalevskaya basically finished her research on the rotation problem in 1886; then she worked for an article for the Bordin prize,* to be awarded in 1888. Her paper [7] was published in 1889 after she had received the prize, which was increased because of the importance of her results. Kovalevskaya also received a prize from the Swedish Academy for her article [8].

We shall touch briefly on the other works of Kovalevskaya.

In [2] Kovalevskaya showed complete mastery of the theory of Abelian integrals and great skill in computation.

In [3] Kovalevskaya studied Laplace's theory on the form of Saturn's ring. Supposing that the ring is liquid, for a first approximation he found the form to be a skew cross-section of an elliptical ring. Kovalevskaya, using a series expansion, found an egg-shaped form as a second approximation. Then astronomers discovered that Saturn's ring consists of solids, and hence Laplace's theory is not applicable. However, Poincaré noted that in one of his investigations he applied Kovalevskaya's power series method.

In [4], [5] and [6] (the latter two articles having the same content) Kovalevskaya studied the problem of light refraction in crystals. Later (1916), Volterra discovered that Kovalevskaya had repeated the mistake of Lamé, who had also worked on this problem. But [4] was still a valuable paper because it contained an exposition of Weierstrass's

---

*Charles L. Bordin was a notary who bequeathed a considerable sum for awards for scientific work, to be granted by the Paris Academy of Sciences.

theory for integrating certain partial differential equations—Weierstrass himself often delayed with the publication of his research.

Finally, in [10] Kovalevskaya gives a new, simpler proof of Bruns' theorem on a property of the potential function of a homogeneous body.

Of course, the main works of Kovalevskaya are the two papers that were discussed in more detail above. They have a great and enduring value.

Translated from the Russian by Neal Koblitz

Literature

1. S. Kowalevsky, "Zur Theorie der partiellen Differentialgleichungen." *Journal für die reine und angewandte Mathematik,* 80 (1875), 1–32.
2. ———. "Über die Reduction einer bestimmten Klasse von Abel'scher Integrale 3-en Ranges auf elliptische Integrale." *Acta Mathematica* 4 (1884), 393–414.
3. ———. "Zusätze und Bermerkungen zu Laplace's Untersuchung über die Gestalt des Saturnringes." *Astronomische Nachrichten* 111 (1885), 37–48.
4. ———. "Über die Brechung des Lichtes in crystallinischen Mitteln." *Acta Mathematica* 6 (1883), 249–304.
5. ———. "Sur la propagation de la lumière dans un milieu cristallisé. *Comptes rendus Acad. Sc.* 98 (1884), 356–357.
6. ———. "Om ljusets fortplanting uti ett kristalliniskt medium." *Ofversigt af Kongl. Vetenskaps-Akademiens Förhandlinger* 41 (1884), 119–121.
7. ———. "Sur le problème de la rotation d'un corps solide autour d'un point fixe." *Acta Mathematica* 12 (1889), H.2, 177–232.
8. ———. "Sur une propriété du système d'équations differentielles qui definit la rotation d'un corps solide autour d'un point fixe." *Acta Mathematica* 14 (1890), 81–93.
9. ———. "Memoire sur un cas particulier du problème de la rotation d'un corps pesant autour d'un point fixe, où l'intégration s'effectue à l'aide de fonctions ultraelliptiques du temps. *Mémoires présentés par divers savants à l'Academie des Sciences de l'Institut National de France*, Paris, 31 (1890), 1–62.
10. ———. "Sur un théorème de M. Bruns." *Acta Mathematica* 15 (1891), 45–52.
11. S. V. Kovalevskaya, *Nauchnye raboty* (Scientific Works). Ed. by P. Y. Polubarinova-Kochina. AN SSSR, Moscow, 1948.
12. I. G. Petrovsky, *Lektsii ob uravneniyakh s chastnymi proizvodnymi* (Lectures on Partial Differential Equations). Gostekhizdat, Moscow, 1950. [Engl. Translation: Partial Differential Equations. Philadelphia. 1967]
13. O. A. Oleinik, *Teorema S. V. Kovalevskoy i ee rol' v sovremennoy teorii uravnenii s chastnymi proizvodnymi* (Kovalevskaya's Theorem and its Role in the Modern Theory of Partial Differential Equations). *Matematika v shkole,* Moscow, 1976.
14. V. V. Golubev, *Lektsii po integrirovaniyu uravnenii dvizheniya tyazhelogo tverdogo tela okolo nepodvizhnoy tochki* (Lectures on Integrating Equations of Motion of a Heavy Rigid Body Near a Fixed Point). Gostekhizdat, Moscow, 1953.

15. P. V. Kharlamov, "Dvizheniye giroskopa S.V. Kovalevskoy v sluchaye B. K. Mlodzeevskogo" (The Motion of S. V. Kovalevskaya's Gyroscope in the Case of B. K. Mlodzeevsky). In the book *Mekhanika tverdogo tela* (Mechanics of a Rigid Body), 7. Naukova Dumka, Kiev, 1975.
16. *Piśma K. T. G. Vaiershtrassa k S. V. Kovalevskoy* (Letters of K. T. G. Weierstrass to S. V. Kovalevskaya), ed. P.Y. Polubarinova-Kochina, Moscow, 1973. A bilingual edition in Russian and German.

A Note on the General Literature

A great deal of memoiristic material has been produced by Kovalevskaya's contemporaries, but unfortunately none of it exists in English, with the exception of the reminiscences of Leffler (Sonya Kovalevsky. A Biography by Anna Carlotta Leffler, Duchess of Cajanello. Translated from the Swedish by Clive Bayley, New York, The Century Company, 1895). Leffler makes many errors of fact, but her psychological perceptions are sometimes acute and the book is worth reading. The same problem afflicts the secondary literature. It is almost entirely in Russian, except for a few items in Polish, French or German. All that we have in English are some items in biographical dictionaries such as the one published by the American Council of Learned Societies, which contains articles on all three Kovalevskys, Sofya, Vladimir and Alexander. Bell's *Men of Mathematics* also contains some pages on Kovalevskaya.

Courant in Göttingen and New York

The Story of an Improbable Mathematician
By **C. Reid**
1976 iv, 314p. 36 illus. cloth

From the reviews:

"This story of the German-American mathematician, Richard Courant (1888-1972), founder of mathematics institutes at both Göttingen and New York University, belongs in every high school and university library and on the night table of anyone interested in mathematics .... clearly narrated ... many insights .... a genuine contribution to the history of science. Highest recommendation."

AAAS Science Books and Films

"Reid has been where historians have not, in the private papers of Courant ... and she has used them sensibly .... She has done the same with the many interviews she conducted with Courant's friends and colleagues. The result is excellent journalism, informative and engaging...an appropriate successor to Reid's well-regarded book on Hilbert."

Science

"...a story of great mathematicians and their achievements, of practical successes and failures, and of human perfidy and generosity .... this is one of the still too rare occasions in which mathematicians are shown as frail, flesh-and-blood creatures ... a very worthwhile book."

Choice

"...an excellent, well-balanced study of a complex and controversial person. Highly recommended for science libraries and for large public libraries."

Library Journal